玛莎·施瓦茨 &
天津大学建筑学院风景园林系

联合设计教学实验

主编 曹 磊

编委 张春彦 刘彤彤 刘庭风 何 捷
张秦英 胡一可 赵 迪 王 苗

江苏凤凰科学技术出版社

项目借助天津大学建筑学院创新引智培育基地的平台，得到了学院领导张颀、张玉坤、宋昆、孔宇航、袁大昌、刘彤彤、许蓁等老师的大力支持。

交流活动简介

2013 年 4 月，哈佛大学终身教授、国际著名景观设计师玛莎・施瓦茨（Martha Schwartz）女士，以及 Martha Schwartz Partners 事务所合伙人马库斯・詹斯奇（Markus Jatsch）先生，受风景园林研究所邀请来我院就景观设计和创作中的可持续性展开交流。期间，曹磊、盛强、张春彦、胡一可、张昕楠等教师同玛莎・施瓦茨团队邀请天津风景园林学会、园林规划设计院、天大设计总院等一线工作者参与主题讲座与研讨，力求达到研究课题理论与实践的统一。双方还共同组织了为期三周的设计教学实验，旨在提升学生对景观艺术品质的认识。

本次玛莎・施瓦茨与天津大学建筑学院风景园林系的联合设计以艺术为切入点，超越风格和形式语言的借鉴，训练一种开放式的，但又不失逻辑性的设计思维方式。设计从参观 798 的作品开始，自选感觉印象深刻的艺术品并对其所表达的理念及手法进行讨论。“回到你选的作品中去，思考它何以能打动你”是联合设计前半部分的主要内容，成果为雕塑模型。而后半部分，将第一部分所选取的艺术作品进行转译，最终形成景观设计方案，其中涉及景观设计语汇的讨论。概念必须通过作品的材料、特性、组合等能被观者直接感受到的东西进行传达，让学生放弃惯用的从功能、流线和行为分析来切入设计的方式，而从一件与基地完全不相关的，但能打动内心的艺术品出发来开启设计。设计过程中也不强调功能理性对形式的影响，更关注对感受的传达是否直接、有效。

玛莎·施瓦茨

博士

哈佛大学终身教授

美国注册景观建筑师

英国皇家建筑协会会员

景观作为文化的人工制品，应该用现代的材料制造，并且反映现代社会的需要和价值，玛莎·施瓦茨一向以不走寻常路和挑战传统的设计手法而享誉国际景观建筑界。她认为用低档的材料创造不平凡的效果不再仅仅是一个选择，而是必要的。她对“设计和创作的可持续性”有着独到见解。

玛莎·施瓦茨女士作为景观建筑师和艺术家从业 30 多年。她在美国马萨诸塞州的剑桥和英国伦敦都拥有自己的设计事务所。她的作品充满着独特的艺术气息，在全球享有盛誉，并且已经赢得众多的国际设计大奖。2008 年前任伦敦市长肯·利文斯通任命玛莎·施瓦茨女士为 " 为伦敦设计 " 顾问团委员，与多位世界顶级著名建筑、规划和景观大师们一起为伦敦的公共空间建设提供专业意见和服务，努力为 2012 年伦敦奥运会创造一个美丽的城市面貌。自从 1987 年以来，玛莎·施瓦茨女士一直在美国哈佛大学研究生设计学院担任景观设计教授，并且在 2007 年被哈佛大学授予终身教授。在此之前她曾经分别在美国加州伯克利大学、美国罗德岛设计学院、俄亥俄州立大学和澳大利亚墨尔本大学任教。

授课教师

玛莎·施瓦茨、马库斯·詹斯奇

盛强、何捷、胡一可、张昕楠

评图教师

玛莎·施瓦茨 (Martha Schwartz)　马库斯·詹斯奇 (Markus Jatsch)　罗宾·沃克 (Robin Walker)

曹磊　盛强　张春彦　何捷

胡一可　张昕楠　张秦英　赵迪

序

玛莎·施瓦茨在天津大学建筑学院教授的景观设计课程结束了，师生们的反应不一，反响强烈：有的大赞新颖，也有称许有趣，个别持相反意见的认为生态性考虑较少……以往沉寂的设计教室一下子活跃起来。学生们的思维被调动起来，景观原来可以如此艺术、有趣，设计竟可以这样构思、畅想。以往我们的课程设置包括公共、技术、主干（设计）、人文等，课程体系完备，设计课题类型齐全，但如何教授和启发学生们创新确实是教学的短板，当然这也不仅是我们这个专业存在的问题，整个社会又何尝不是如此——创新动力不足。

玛莎·施瓦茨的作品和她的教学方法一样，让人耳目一新，她是一位景观设计领域的拓荒者，不断探索景观设计新的观念和形式，希望将艺术思想融入她的作品和创作过程中。玛莎·施瓦茨无疑为我们开启了一个视角，从艺术中去探寻景观设计灵感，创作具有原创性的景观方案。创新是艺术的灵魂，当然也是景观设计的重要价值取向。

笔者曾著《中国当代的艺术观念与景观设计》一书，书中通过对现代艺术的源流、形态创新的特征进行分析总结，寻找当代景观创作的原创动力，丰富景观的形态世界；对后现代艺术的发展脉络以及艺术观念转变的特征与意义进行分析和论述，探求当代景观理论研究和创作观念的思路和灵感，丰富景观的观念世界。其中一些思路和方法与玛莎·施瓦茨的有相似之处，但她无疑走得更远。

玛莎·施瓦茨的景观设计课程教学给学生们最大的收获是让他们的思想插上翅膀去飞翔。她给我们的启示有三：其一，景观设计要不断创新，而不是模仿，这是其重要的价值所在；其二，景观设计要凸显其艺术性，特别是让大众能够读懂和欣赏；其三，景观教学应该有方法可依，不是只可意会，不可言传。她已将创新方法言传给学生们，而且不仅言传，还身教，谢谢玛莎·施瓦茨。

——曹磊

学生作品

VISUAL ILLUSION

赵晶、李相逸、陈永辉
Zhao Jing, Li Xiangyi, Chen Yonghui

Design Diagram

This relief sculpture on the wall reflects the aesthetics of acoustic's ripple shape, which inspires us. We get the art image from the art work and redesign it to get the mutual interference.

At the same time, we set off from the way of thinking and the angle of viewing. In traditional way of design, people always use 2D image to show 3D image effect. By constract, we use the way to create this art work, which means using 2D image to show 3D image effect. From the 2D and 3D transition, we get the visual illusion.

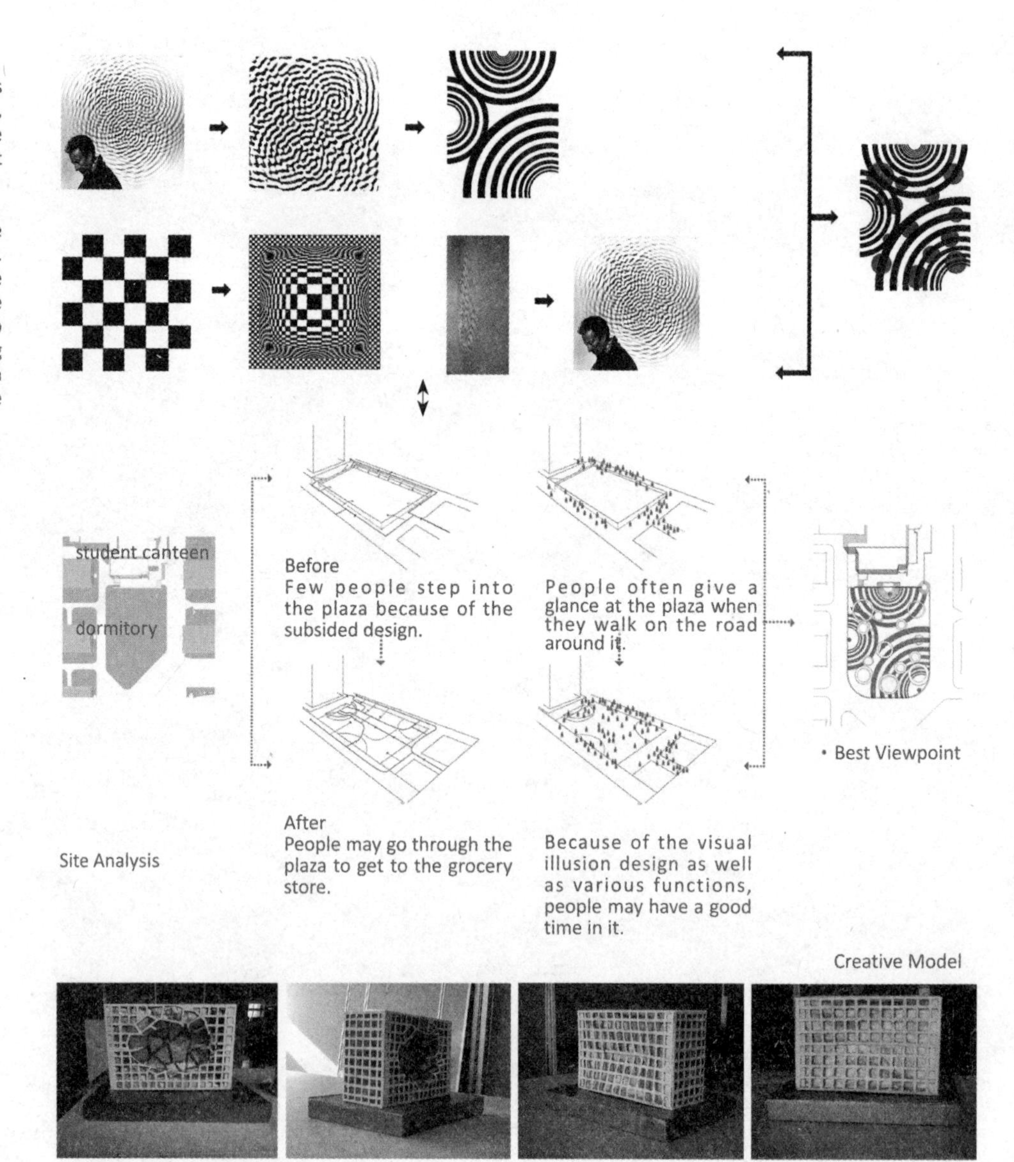

Design Progress

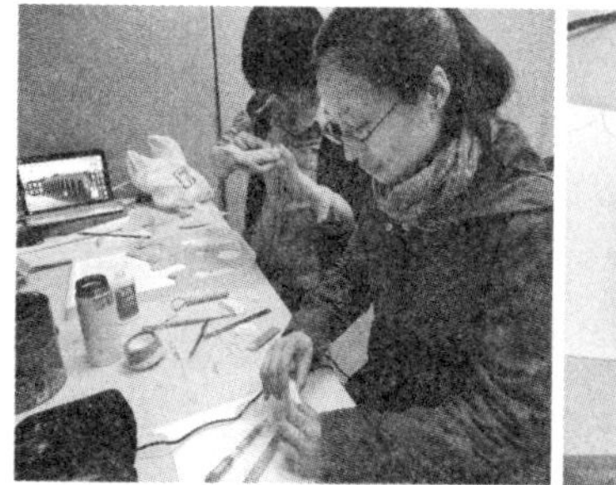

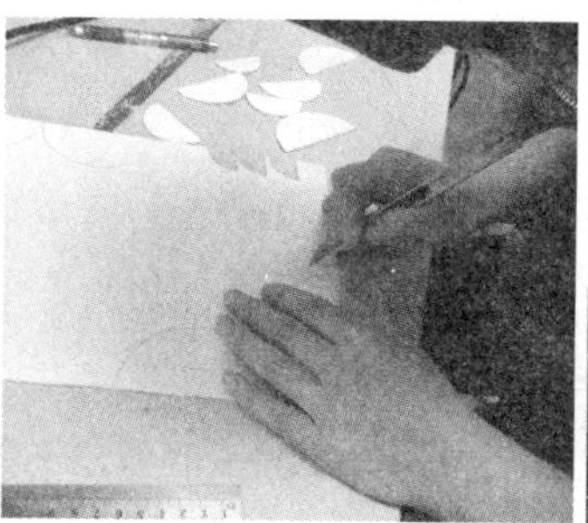

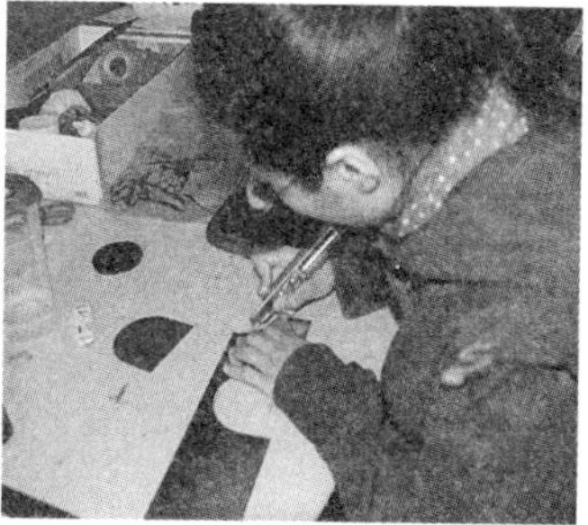

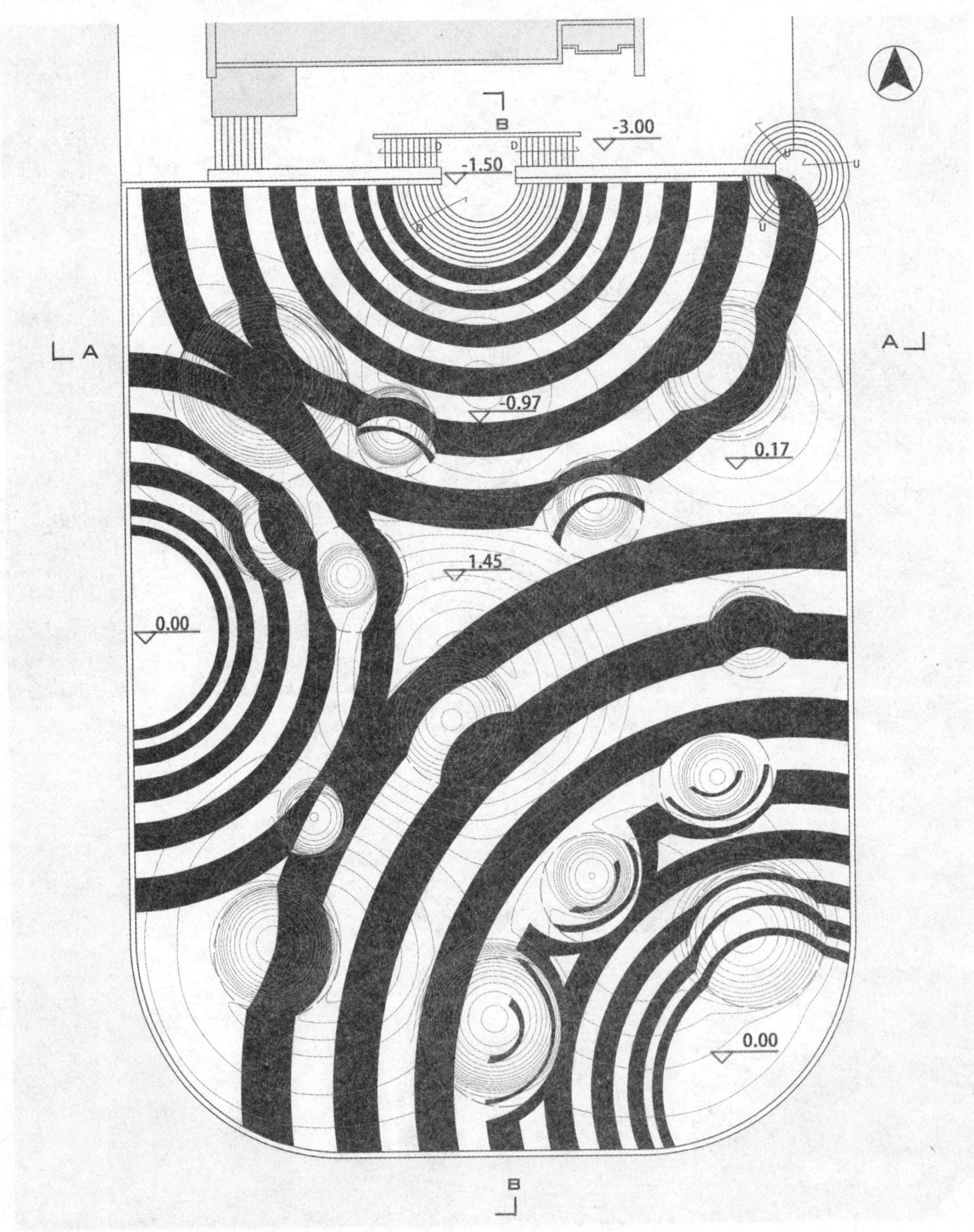

Site Plan

Design Principle

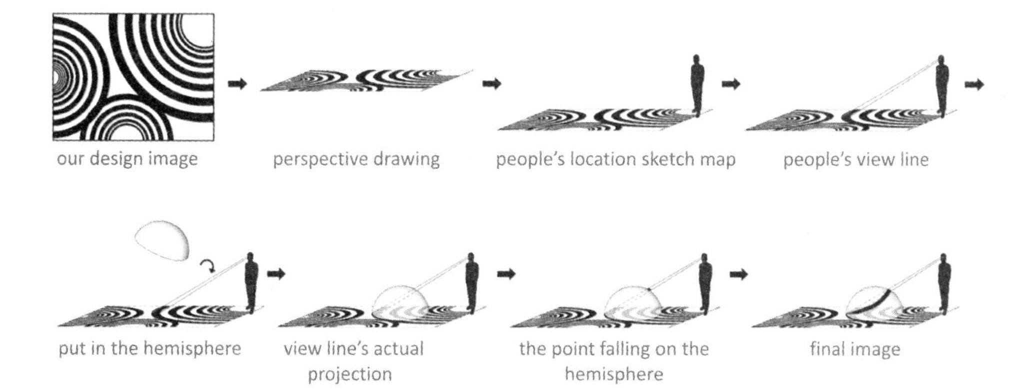

Manual Model Photos

Functional Diagram

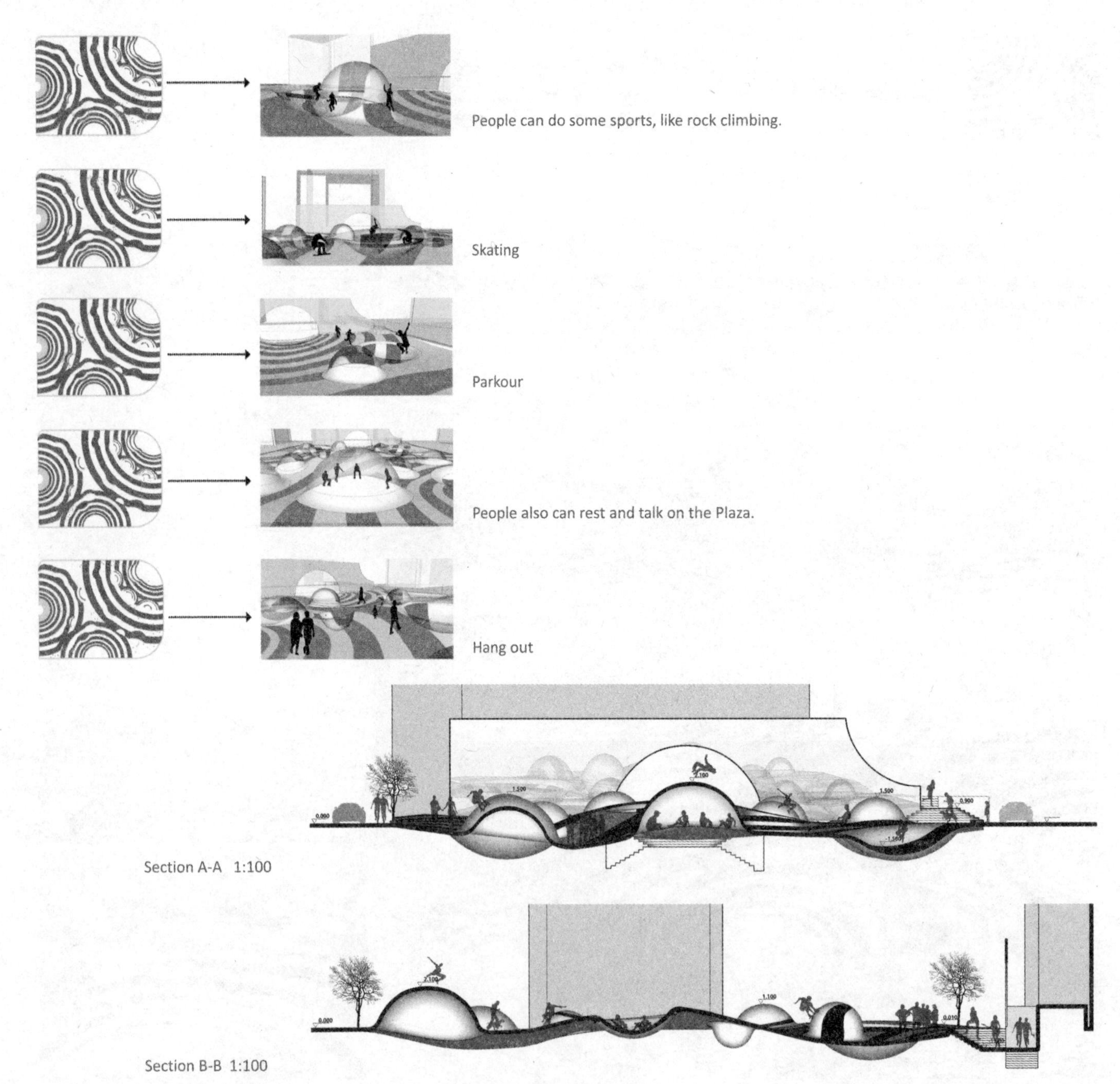

Section A-A 1:100

Section B-B 1:100

Material Intentions

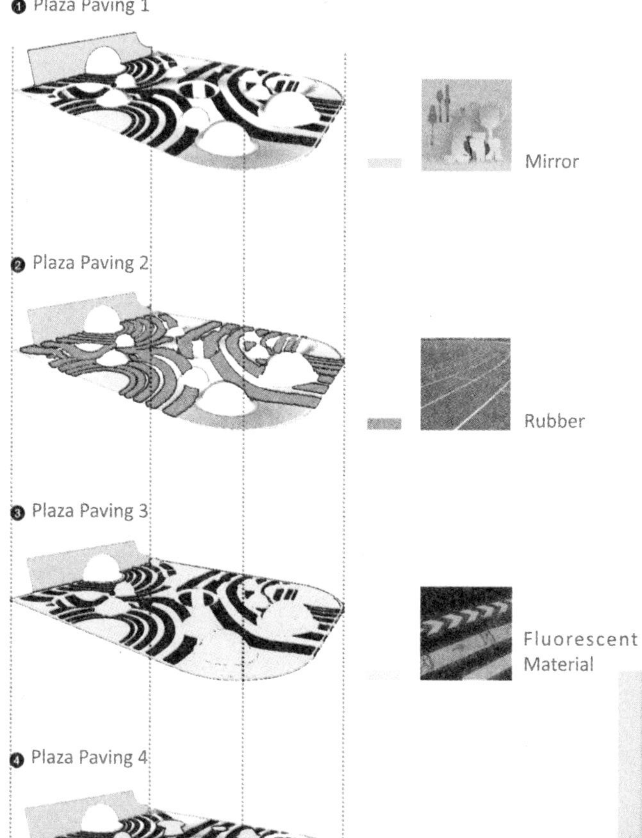

Perspective

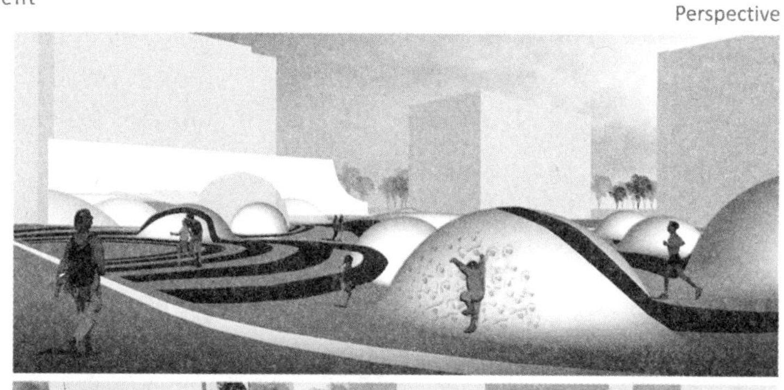

As the main place of activity, the plaza needs to meet the basic demand of doing sports and leisure for people. Therefore, the plaza paving mainly uses the rubber which is the most suitable material for the sports. The mirror at the entrance of the student canteen is good for widening space and makes the visual effect more perfect. In addition, in order to show the interesting effect of visual illusion in the night, part of the space uses fluorescent material to improve night's effect.

Perspective

Bird's Eye View

Mirror 1
Rubber 2
Fluorescent Material 3
1
3
2

IN BETWEEN

郝钰、沈悦、林安冬
Hao Yu, Shen Yue, Lin Andong

What impressed us most is a time-lapse photography. Screenshots of a video clip in art gallery, which we chose as our starting point.

Task A: Sculptural Model

Before the landscape design, we designed a small sculpture model based on 2 kinds of given materials: cardboard and clay which are in contrast with each other: shoebox is hard and flammable, while clay is deformable when heated and nonflammable.

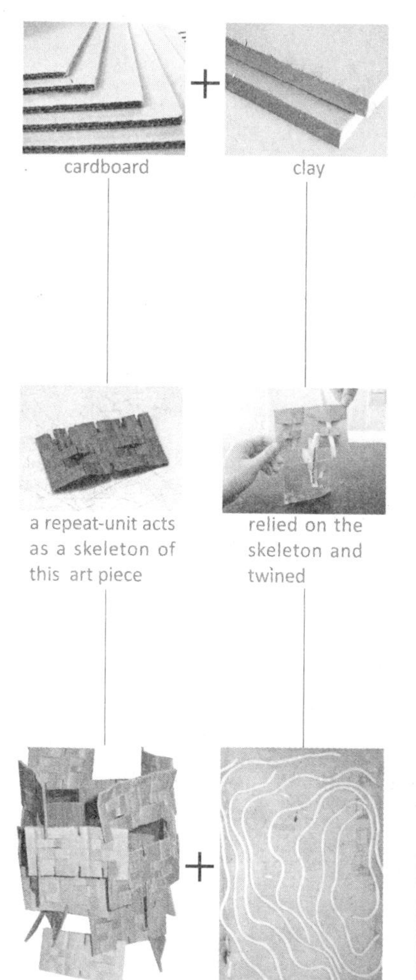

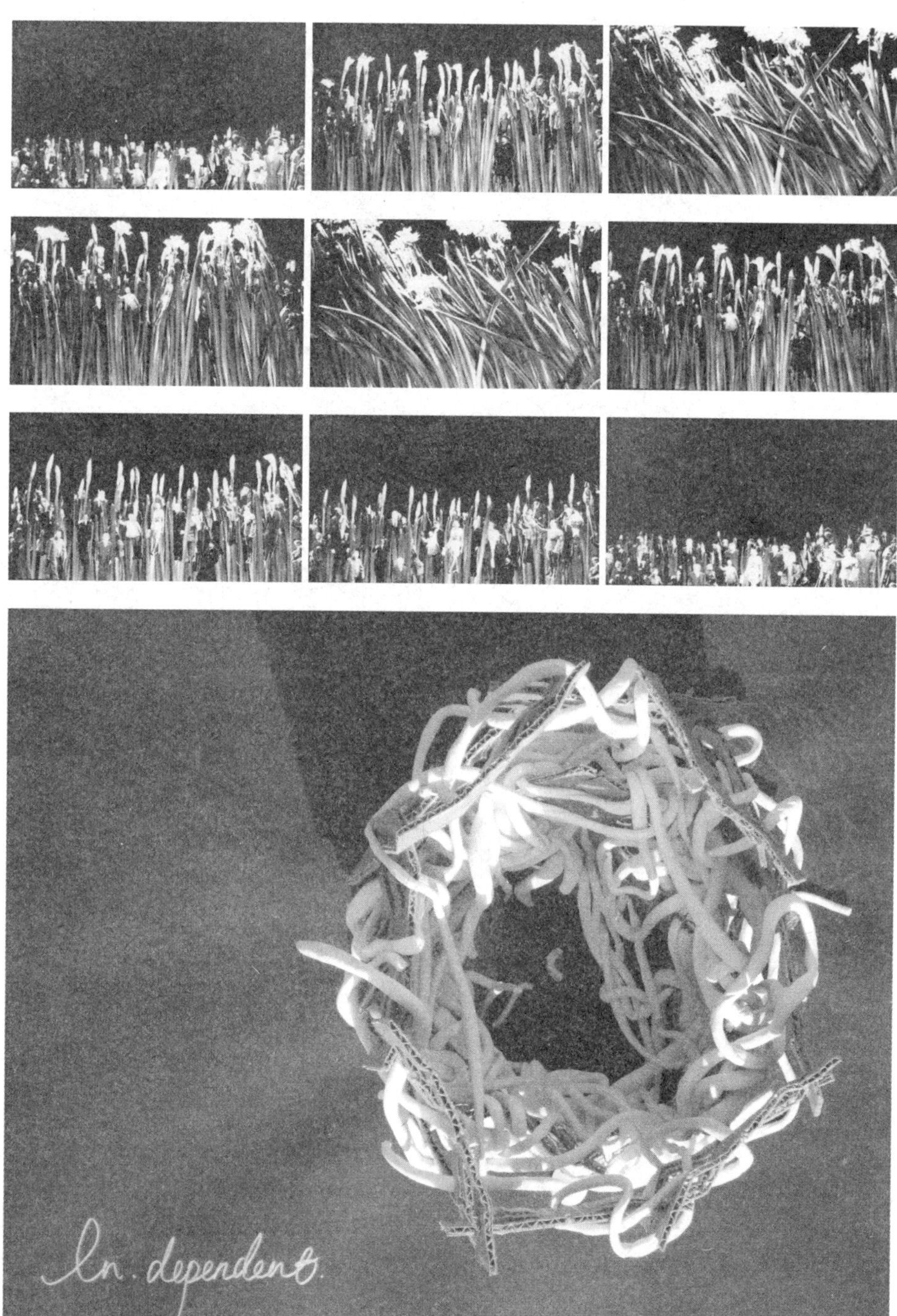

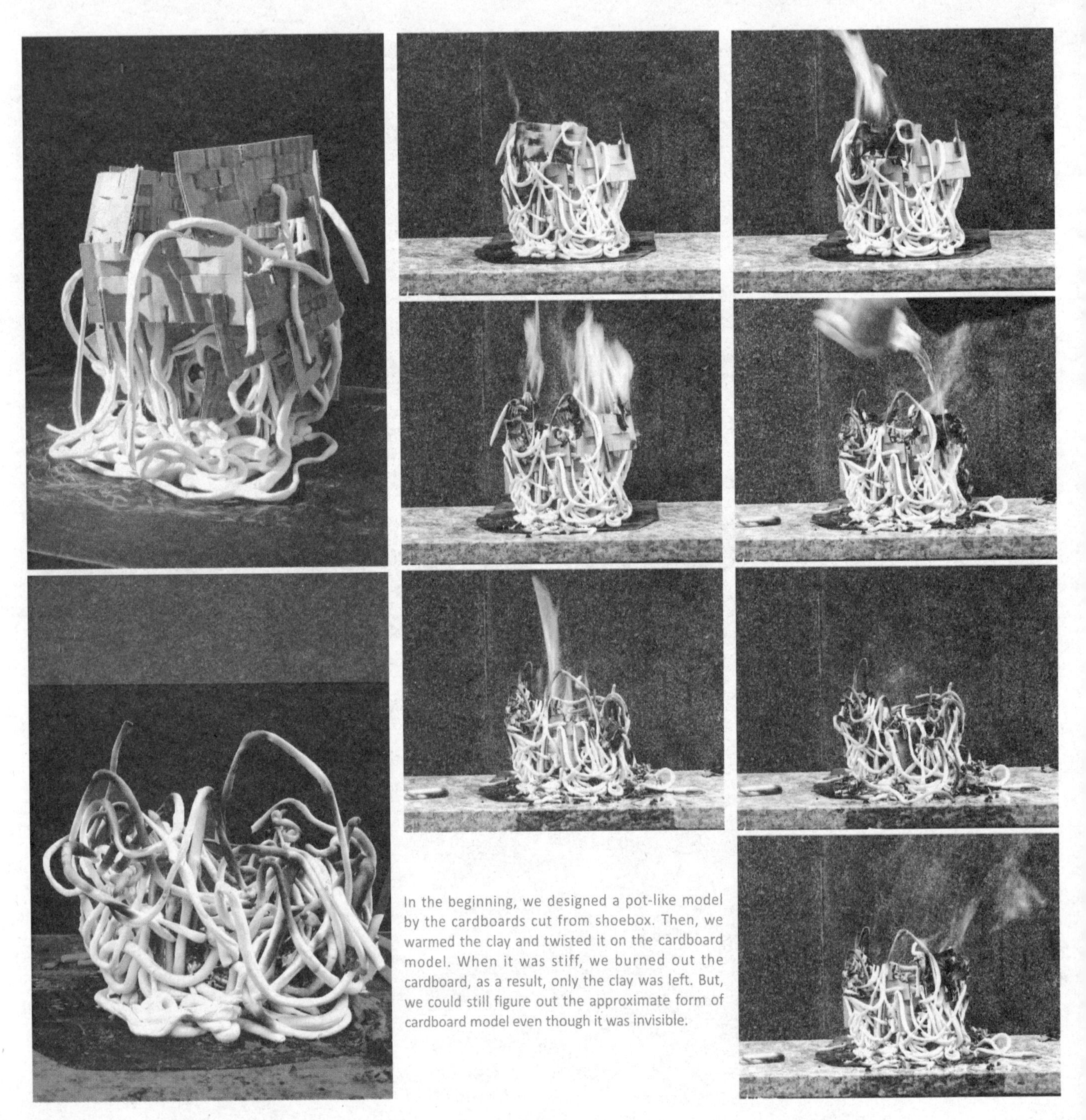

In the beginning, we designed a pot-like model by the cardboards cut from shoebox. Then, we warmed the clay and twisted it on the cardboard model. When it was stiff, we burned out the cardboard, as a result, only the clay was left. But, we could still figure out the approximate form of cardboard model even though it was invisible.

Task B: Landscape Design

The site locates in front of the student canteen in Tianjin University. Beneath the canteen is the grocery store. An array of columns are supporting the daffodil while the passers-by represent the leaders in video.

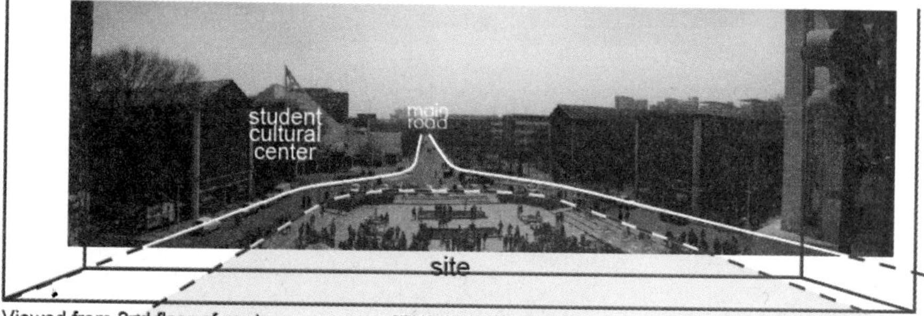

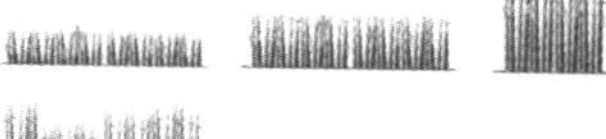
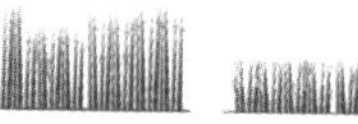

Interpretation

The interaction forms between portraits of leaders and daffodil keeps changing. The leaders are either hidden by daffodil or the reverse. Accordingly, the concept for this design is characterized by this typical relationship.

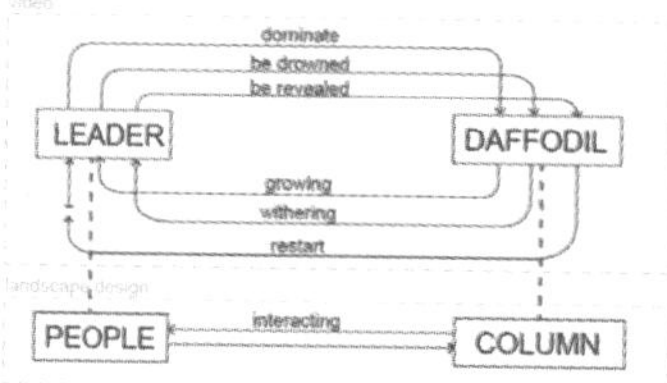

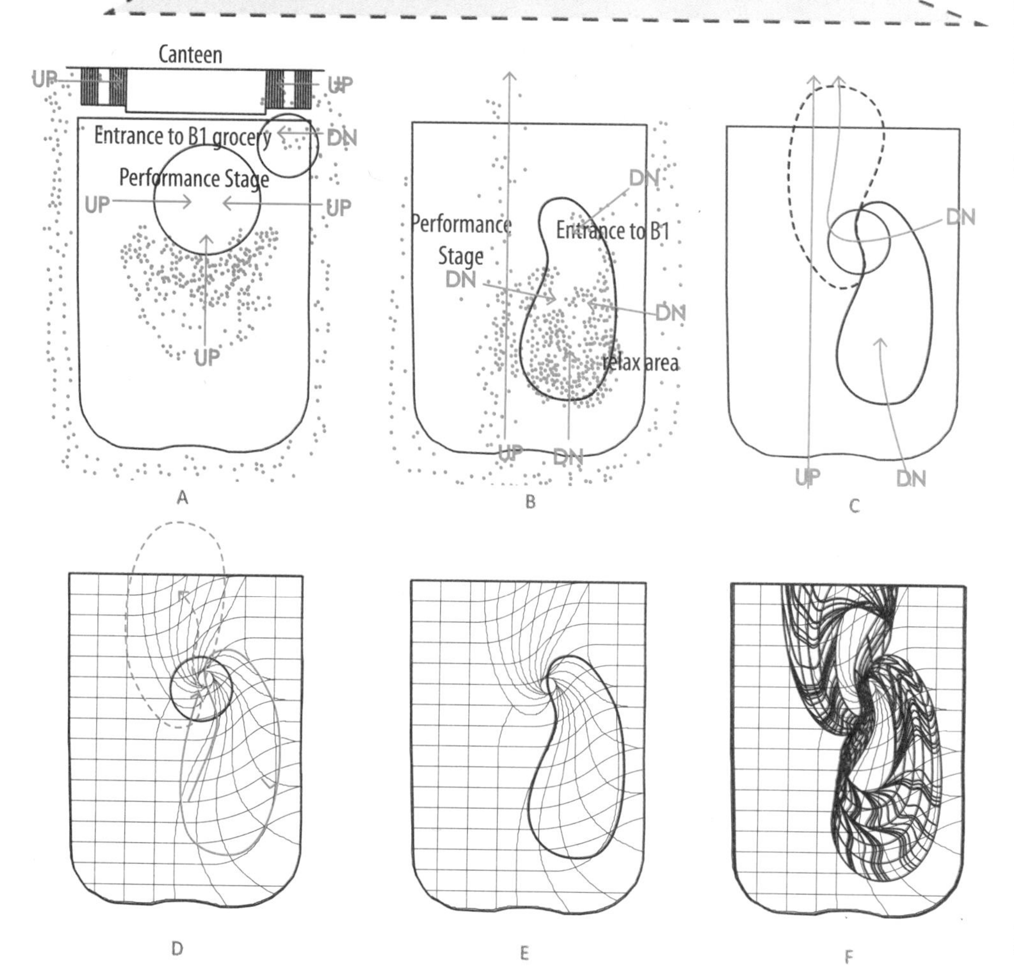

A B C

D E F

A. Current Site
· low efficiency
· 2 main functions: performance + entrance

B. Rebuilt Site
· combination of performance area and entrance

C. Rebuilt Circulation
· 3 routes independent but also connected to each other

D. In and Out
· surface of the stage extending into the B1 entrance area like a blackhole

E. Column Pattern
· the blackhole twists the ordinary grid pattern into typhoon-like shape, offering different activities in such appropriate site condition

F. Modeling

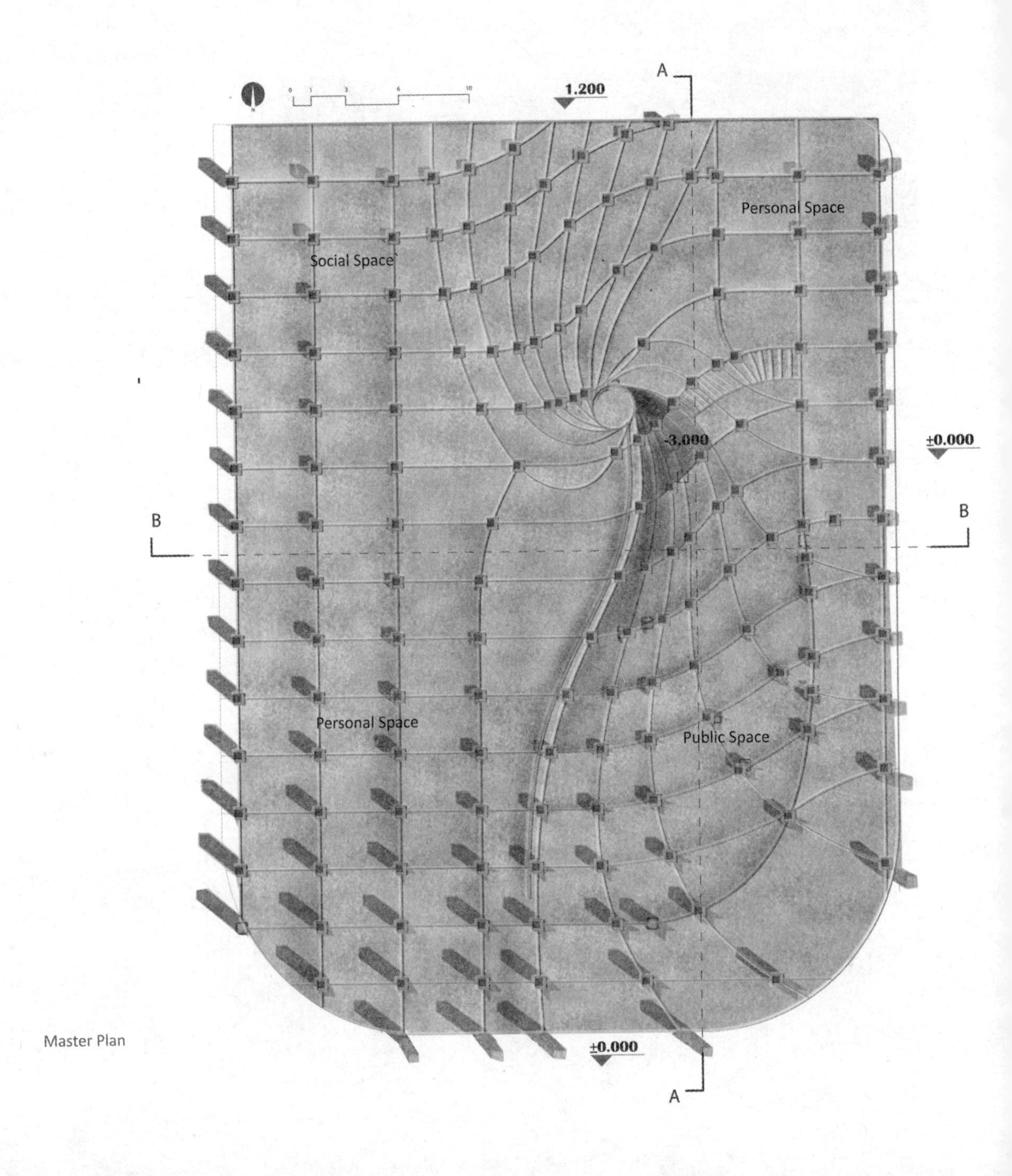
A
1.200
Personal Space
Social Space
-3.000
±0.000
B
B
Personal Space
Public Space
±0.000
A

Master Plan

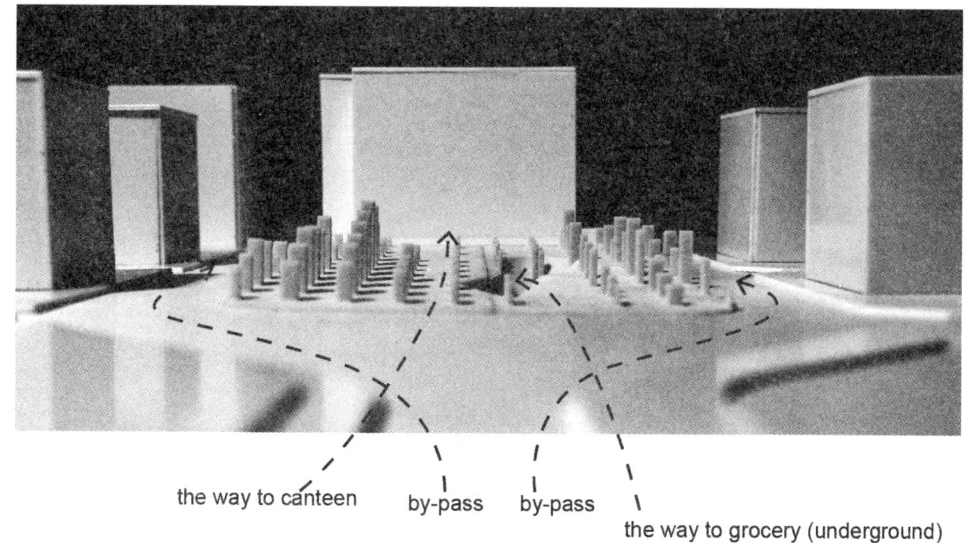

Interaction within Site

We visualized and highlighted the circulation through a subtle slope up to the canteen and a sunken plaza down to the grocery store underground. Furthermore, the special design of columns would increase the possibilities for various interaction.

High fat columns provide a relatively private space.

Short columns provide not only open space but also chair for public activities.

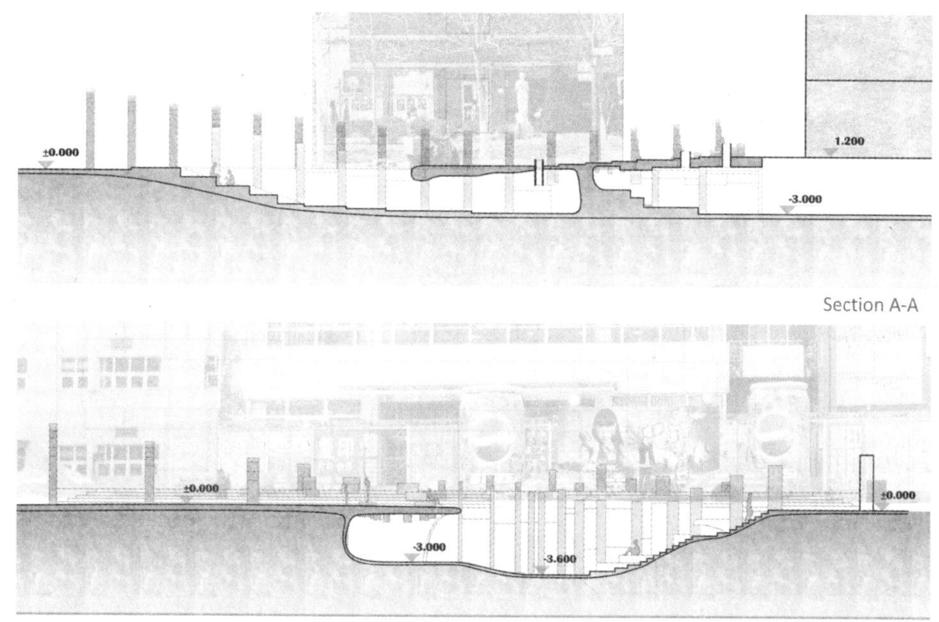

Section A-A

Section B-B

The columns above tunnels become the light wells.

Light wells promote interaction between people aboveground and underground.

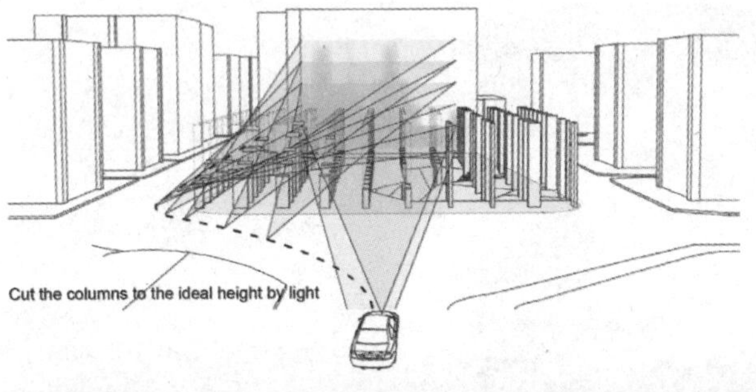

Interaction with Surroundings

The changing position relationship between still columns and walking people corresponds to the scene in the video. Further, the car could project the scene on the wall of the canteen, thus precipitating the site interactions.

The plaza needs to provide a fast path connecting the cateen and the grocery store in the basement.

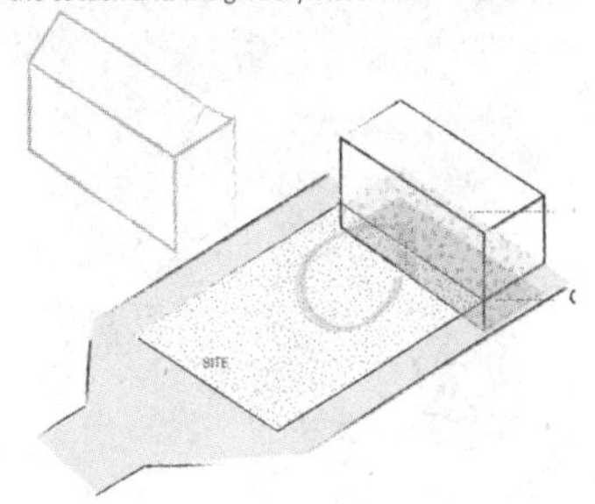

Twisted the grid and shaped the ground act as an implement to attract people to the basement.

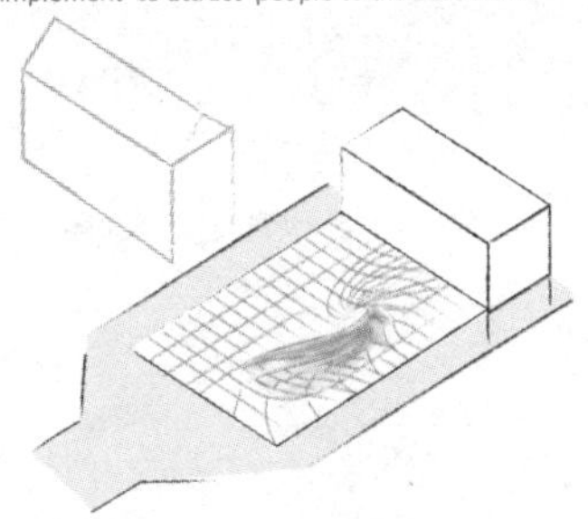

Column is an art element to confine space which can integrate the column into the environment in stainless steel material.

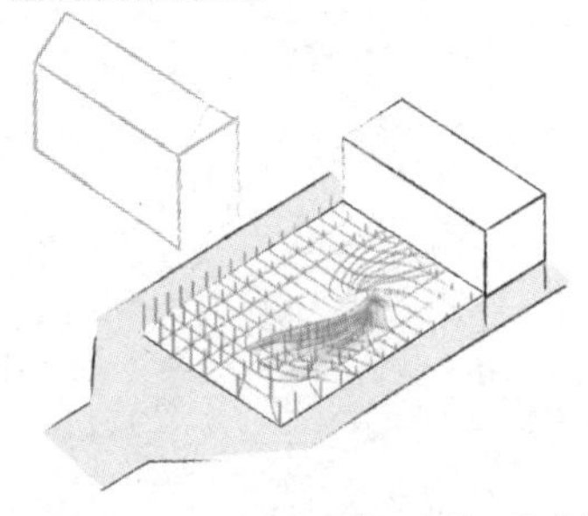

It's like Chinese Shadow Play. This special and joyful experience is not consistent of three dimension but four-light, which makes it artistic.

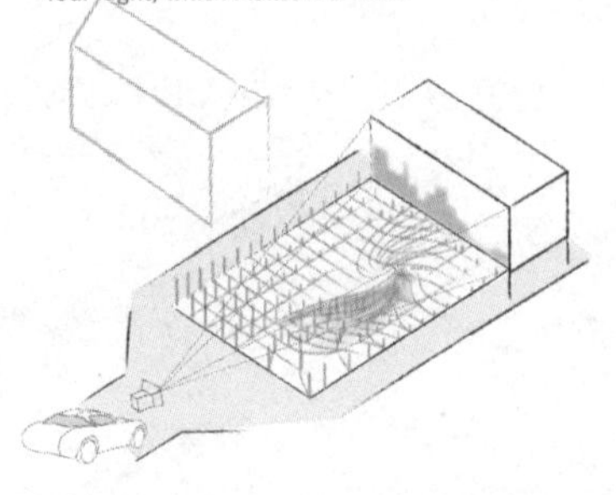

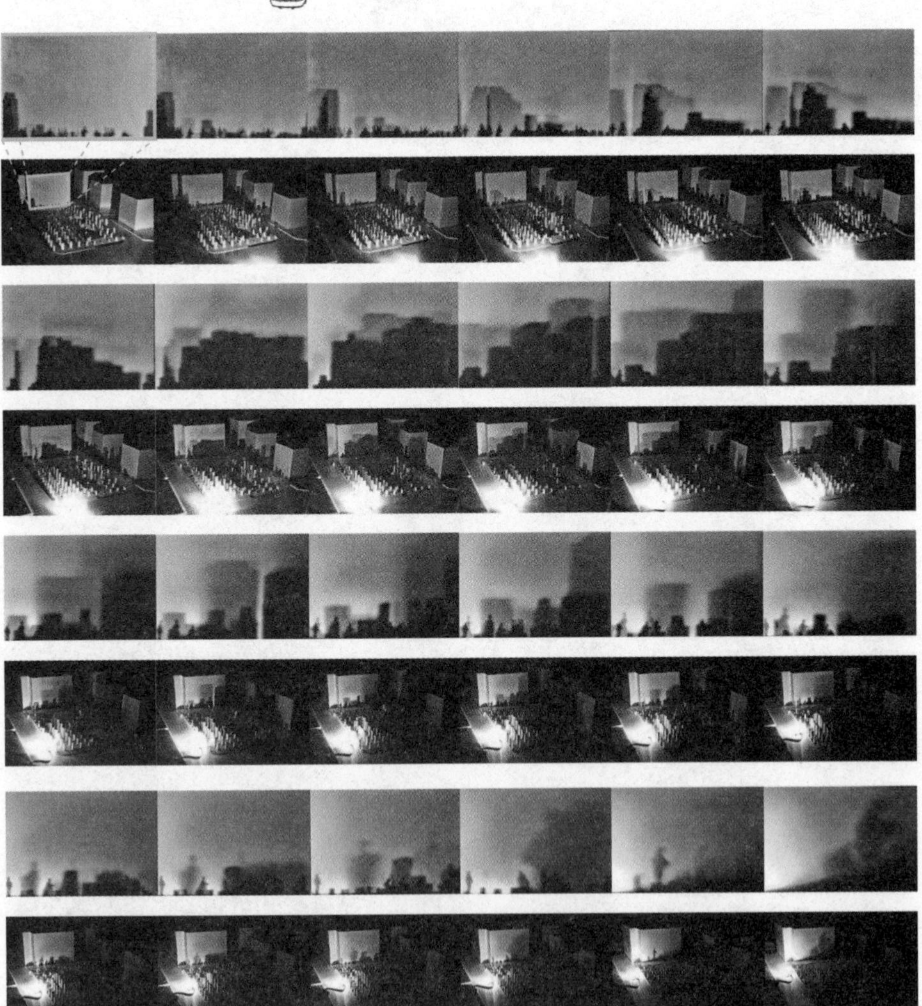

Time-lapse:Two model car's lights are used for simulating the movement of car at night. As expected, the image projected on the wall worked out perfectly except that it's more volatile than the one light analyzed before.

Model

Model of Section

Model of Section

When darkness falls, cars driving by will provide a dynamic light source which project the column forest to the facade of canteen.

The grid scale is 4 m×4 m. It fits very well for both public and private space.

People could even interact with the columns. Some paths can only be viewed from specific points.

A sunken plaza is shaped following the grids, demarcating the underground entrance.

CONTAIN

李璇、陈志飞

Li Xuan, Chen Zhifei

Location Analysis

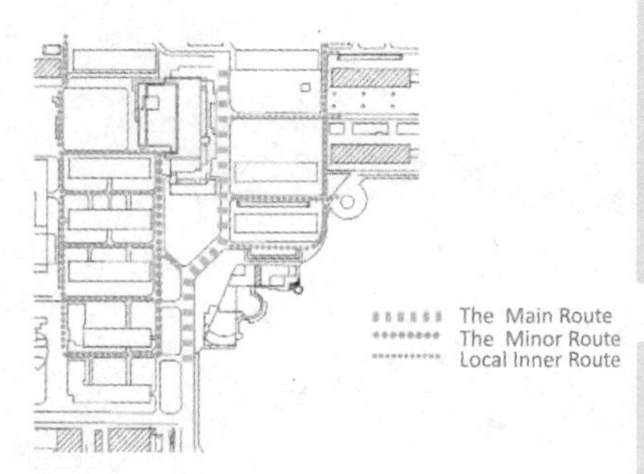

Traffic Analysis

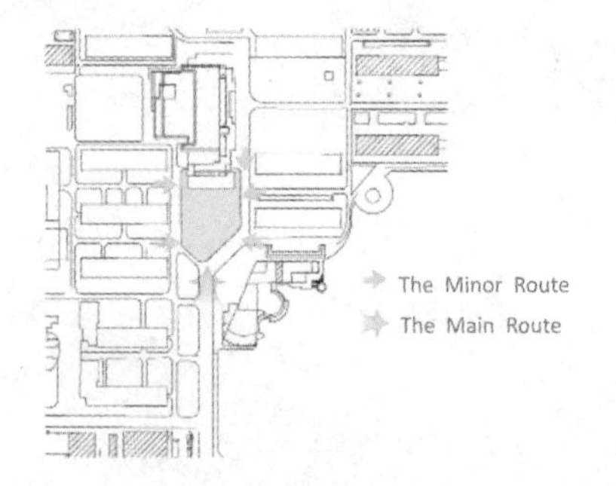

Stream of People Analysis

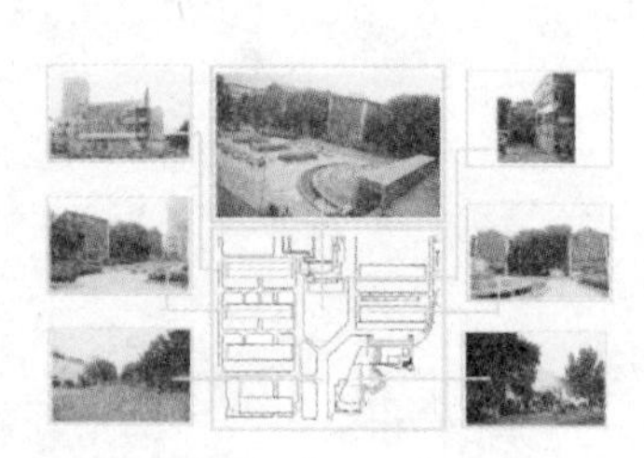

Status Analysis

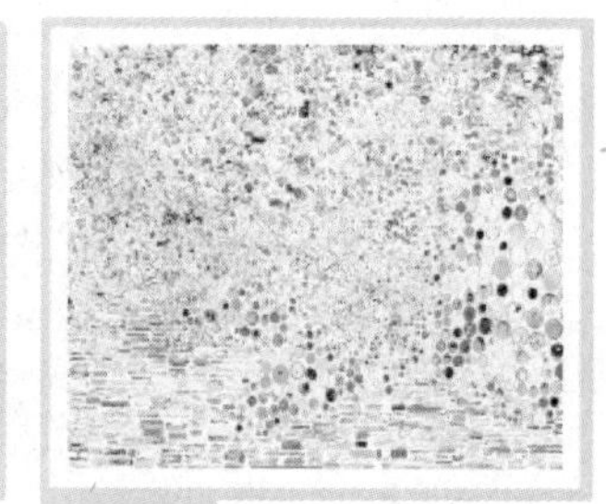

Content

Select the items used in the life. The figure formed by several different groups with representational elements repeated in the form of permutations and combinations.

Structure

Objects disappeared, shapes appeared.

The ideal of the artist:After all the real-life experiences, returning to innocence to the initial state, but more than the newborn which contains more grown up on the wisdom and precipitation.

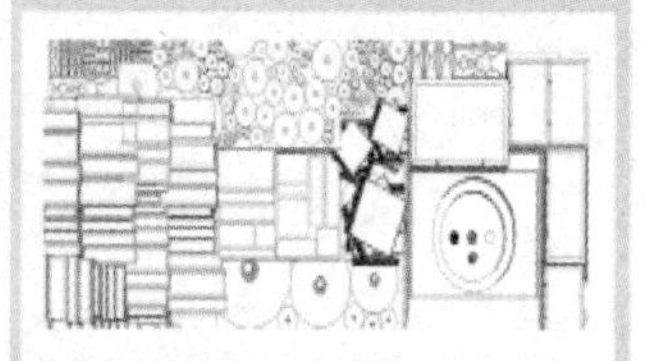

Content

The permutation of articles for daily use

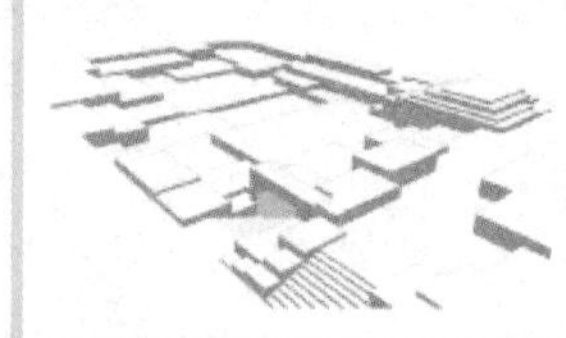

Structure

The permutation of articles for daily use

Materials

The formation of different height of rectangular space is to express the feeling of the fragments.

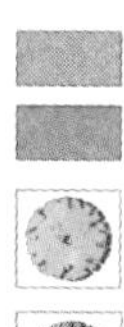

Wood

Grass

Acer Mono

Peach

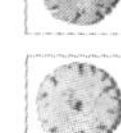

Sophora Japonica

Design elements are chosen in a direct way to respond to the art piece,which also selects the items used in the life. The figure formed by several different groups with representational elements repeated in the form of permutations and combinations, generating intensive and reincarnation feeling. The formation of different height of rectangular space is to express the feeling of the fragments, the fragments in the art piece carry a different figurative elements, but in the landscape design they provide a variety of activities to meet the needs of the students to record the fragments of life. The chosen materials are wood and grass, two different materials on behalf of the life and reincarnation. We try to respond to the art piece through different aspects in direct and indirect ways.

Plan

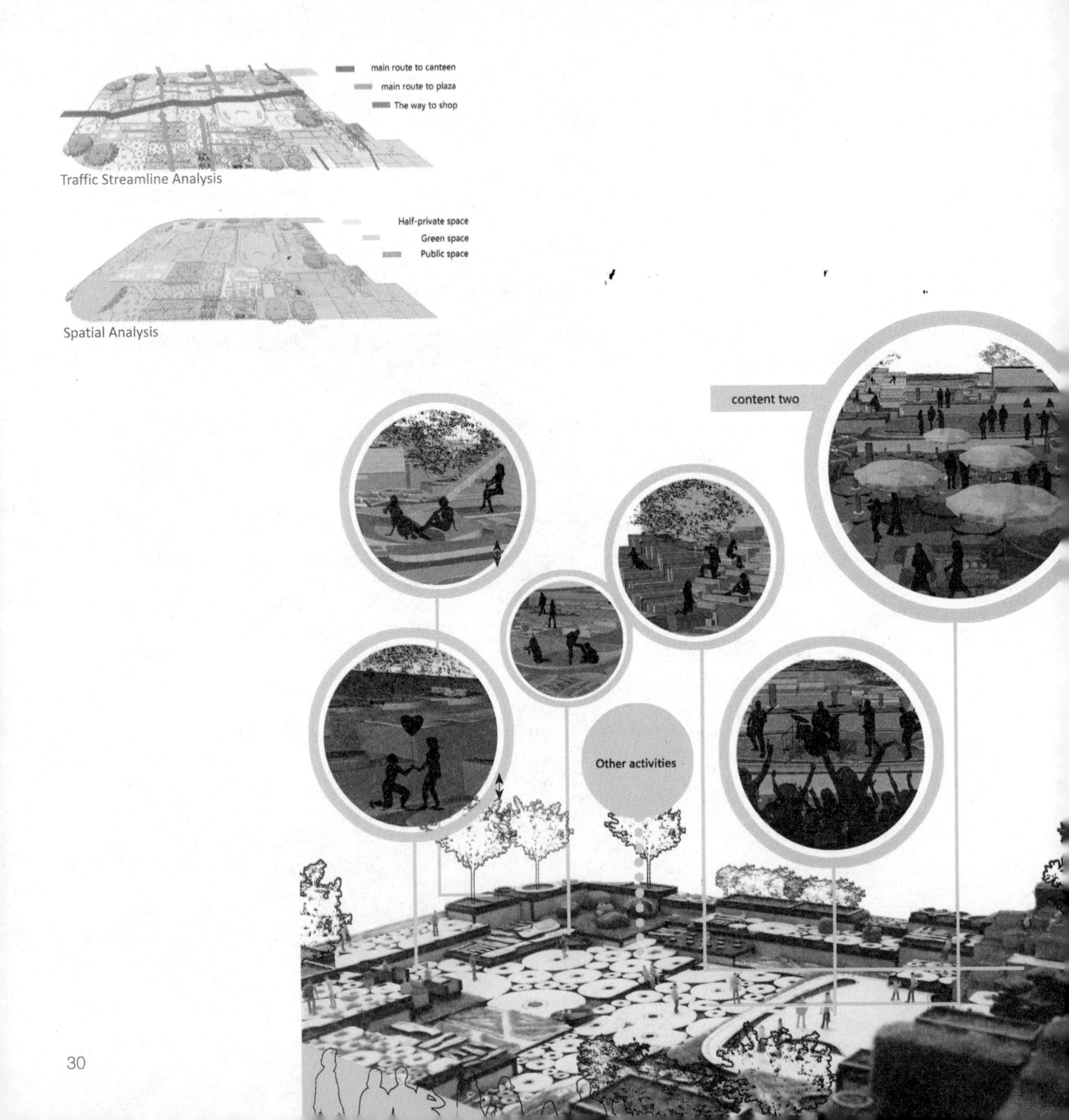
main route to canteen
main route to plaza
The way to shop
Traffic Streamline Analysis
Half-private space
Green space
Public space
Spatial Analysis
content two
Other activities

Overall View

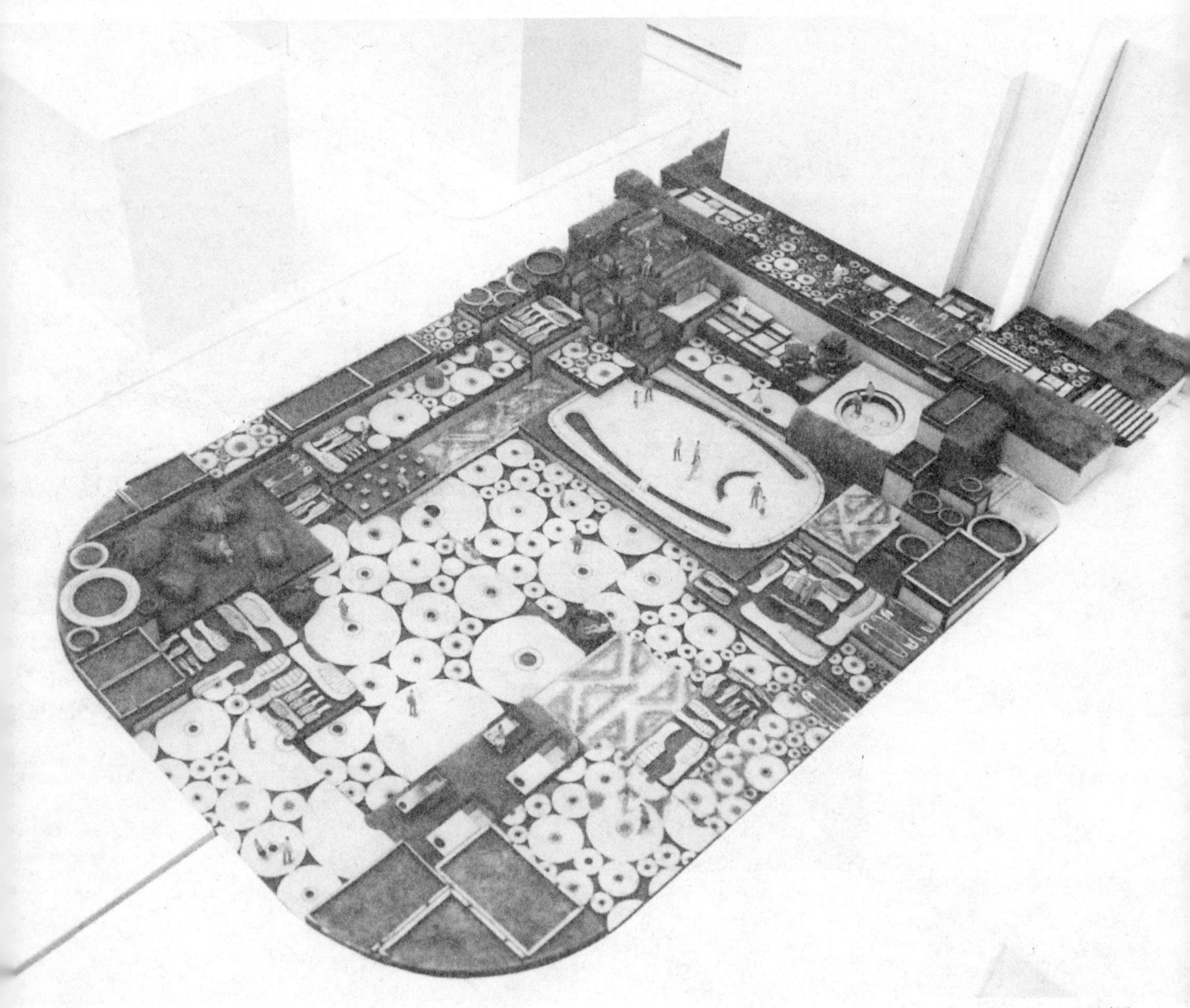

Aerial View

WATER PIECES

王学值、崔海南
Wang Xuezhi, Cui Hainan

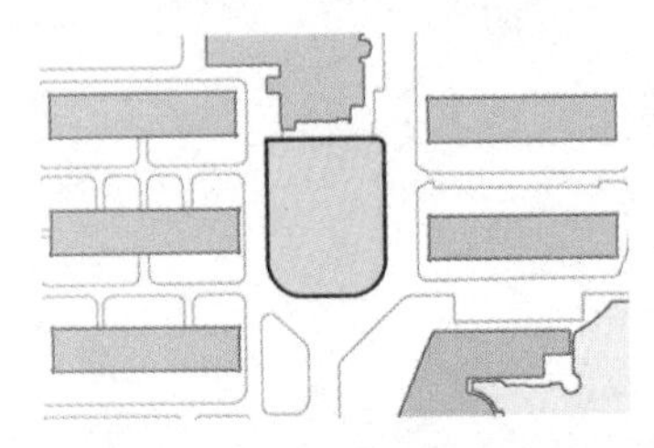

As we have seen, the art elements are in the hierarchy, an industrial, physical framework of artificial and a winding growth is combined with logical organism perfectly.

The organic and inorganic, natural and industrial conflict with each other, rational thinking and emotional thinking have become a source of inspiration for us. We desire to reproduce this conflict in the limited landscape design.

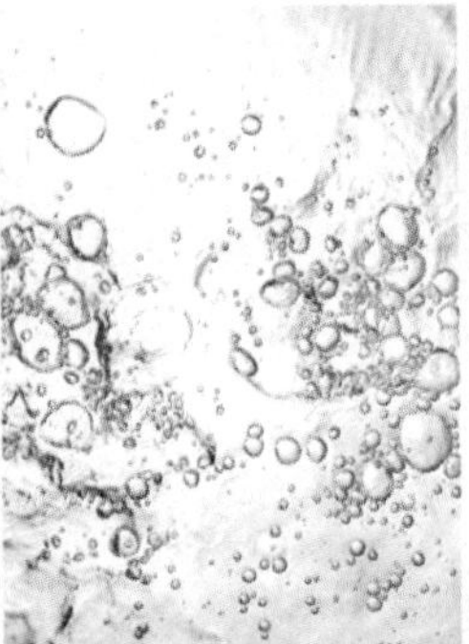

The square base is located between a canteen and dormitory, as an important place for student activities.

Our design especially emphasizes the concept of a kind of avant-courier and good practical function.

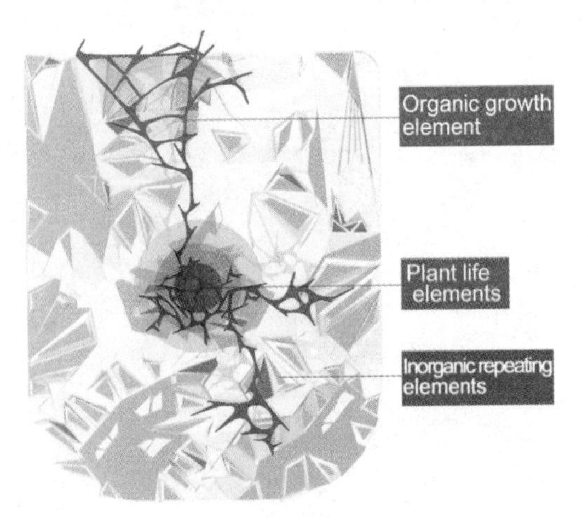

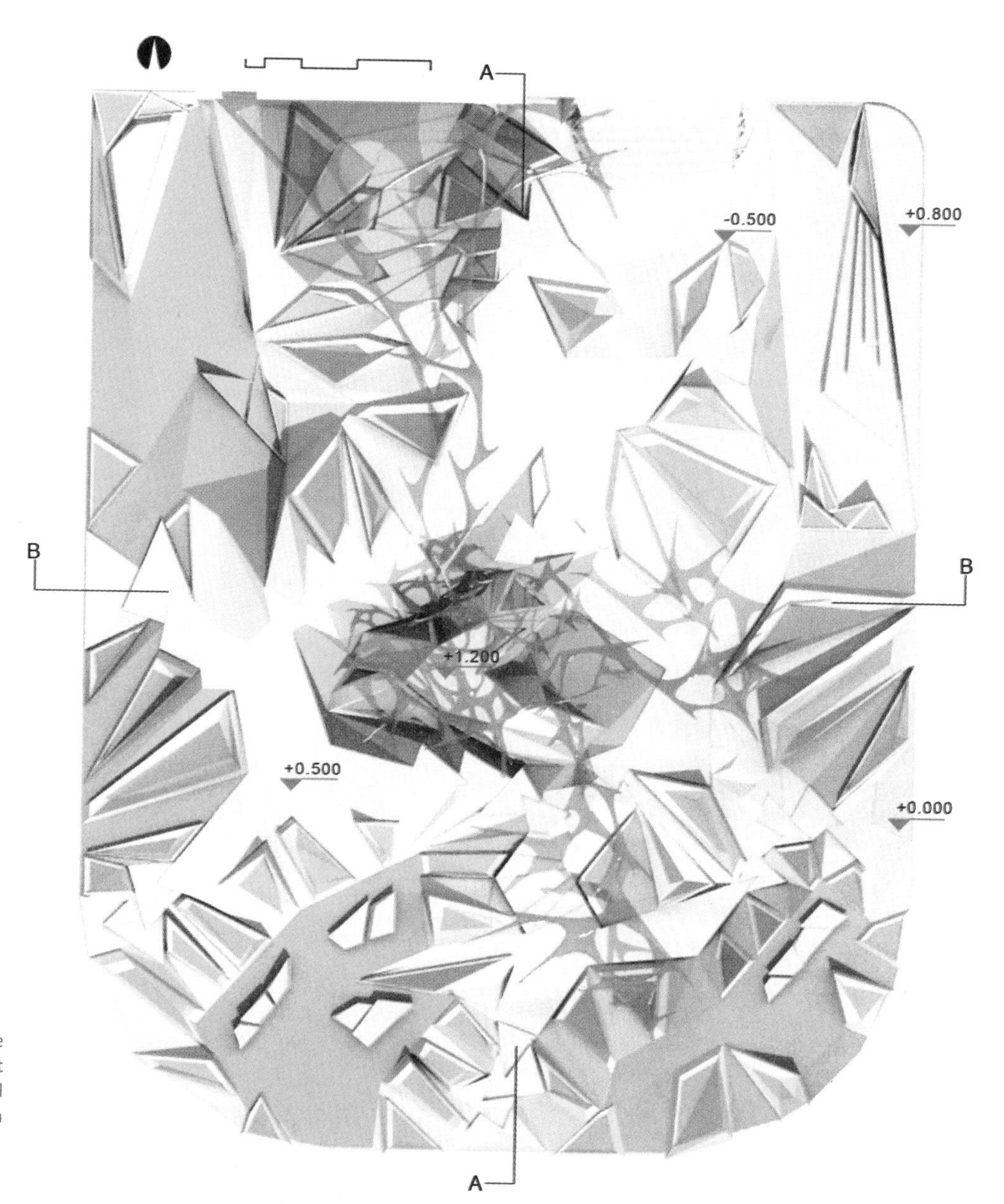

General Layout

From the graph on the surface, the overall feels like broken pieces, in fact it is so produced. We try to use material and color to integrate the broken structure, making it become a basis.

The landscape design of the function analysis, includes its tour line and functional partition. We connected the square on both sides of the road as much as possible and also designed very convenient steps to the upper canteen and the undergroud supermarket.

When reducing unnecessary distance, users will lead themselves into the landscape. We reserved a lot of free green space for the students on the geometry structure around the green space.

We selected a number of surfaces to replace the material for the mirror, cooperating with the organic framework of stainless steel, the image of the user will become a part of the landscape elements. By this means we try to create the atmosphere of user and landscape integration.

Shown in the figure, is the steps of the underground supermarket.

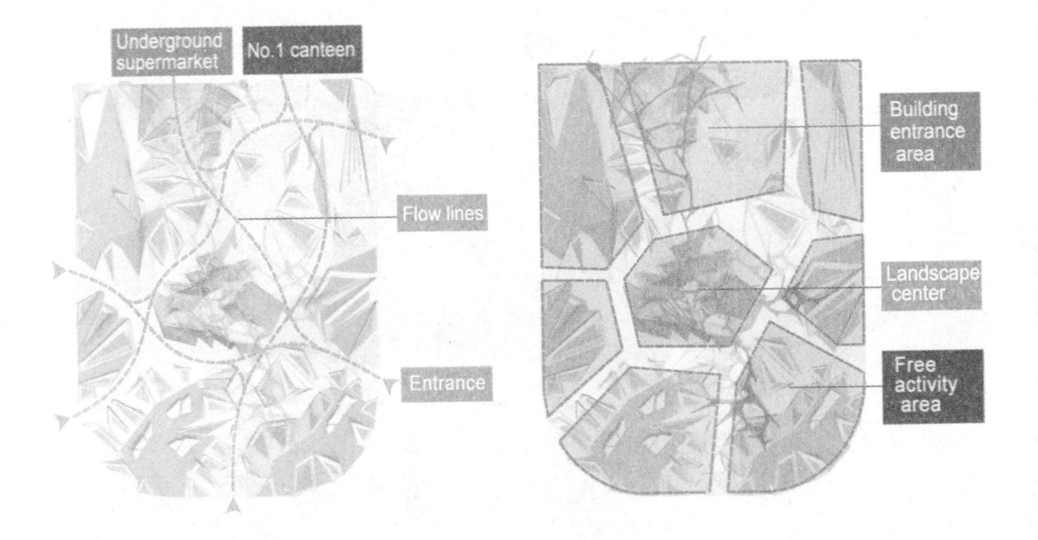

Section A-A

Section B-B

If possible we will plant a tree in the visual center of the landscape, this will be the core of the whole landscape theme:A story about the inorganic soil and organic water gives life together. A living tree is the most suitable for this role.

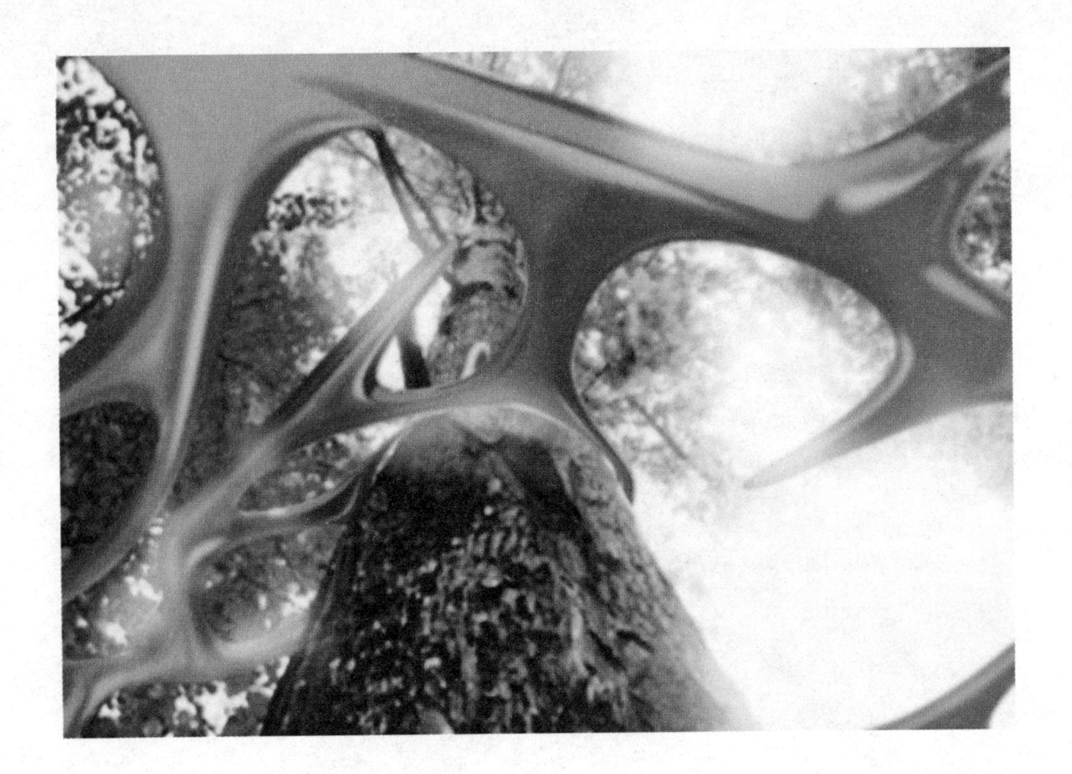

Winding stainless steel organic framework will be surrounded by tall trees, and under it constitutes an interesting space. Projection down the canopy light with stainless steel section can produce interesting visual effects.

On the right side is the mirror structure on the base.

Human element is our priority, we reserved a large amount of activity space for this. Seen from outside, it is very complex, but if you place yourself in it, you will feel it very spacious and comfortable.

FLOWING STAIRS

SHARP AND SOFT

王朝、张茜、边文娟
Wang Zhao, Zhang Qian, Bian Wenjuan

This diagram illustrates the student square in the whole campus environment in the specific location and the main routes to the square.

The three diagrams on the right analyze the formation process of the site. Firstly, according to the main roads, surrounding buildings and stream of people, we generated the main structure in the square. For the need of the function, we made the site sunken to form a stage. Because the 3rd floor of the canteen is the youth centre where students usually gather, this will bring the platform overhead by the structure of columns, eventually leading to the "flowing stairs".

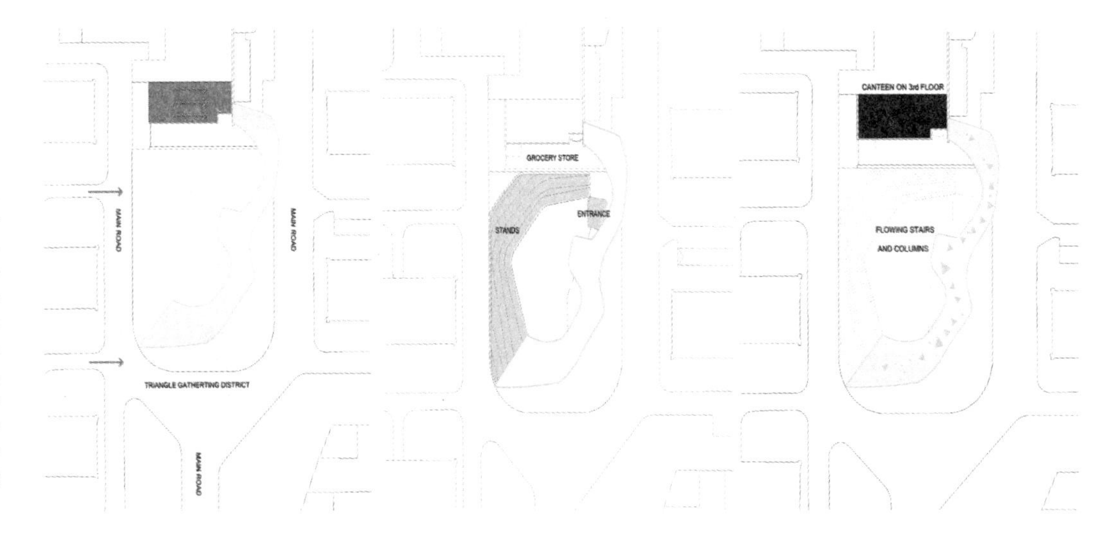

The diagram on the right describes how we approached the sharp pointed things to be soft. Sharp triangles rise up row by row, then by adjusting the elevation angle we tried to form a surface, which is smooth and soft.

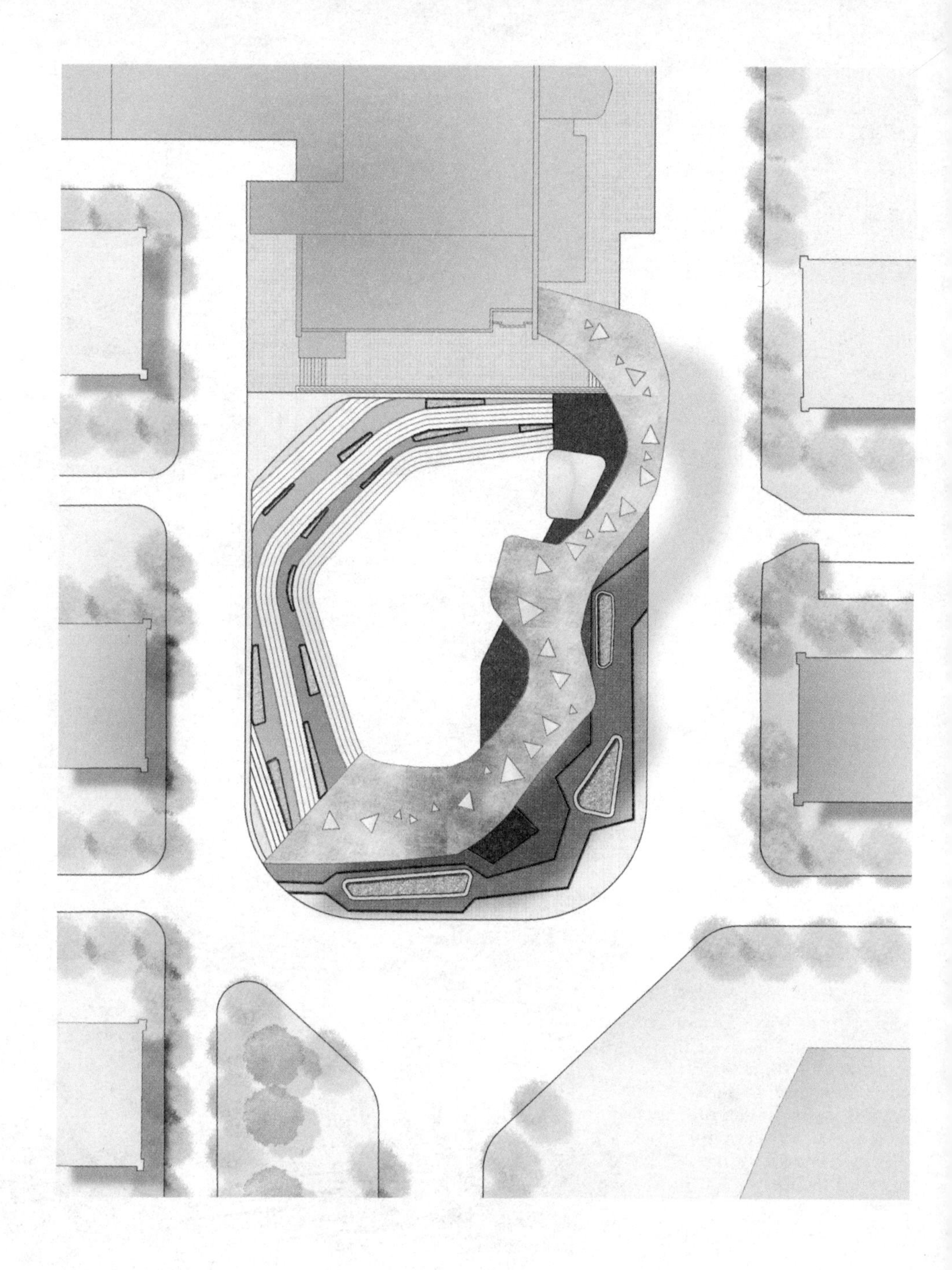

Site Plan

This perspective shows the material and structural details of the flowing stairs.

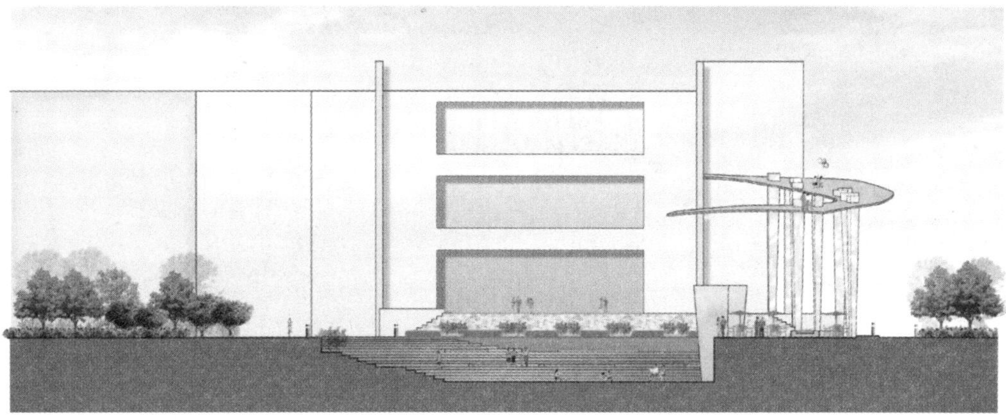

Section

Perspective

SITTING BY

王杨、高哲、张梦蕾
Wang Yang, Gao Zhe, Zhang Menglei

"Bees" designed by Chen Zhe. It records a group of people who tend to harm themselves.

Using sticks and clay to build a sculpture based on the chosen art piece. They are escaping from that little one.

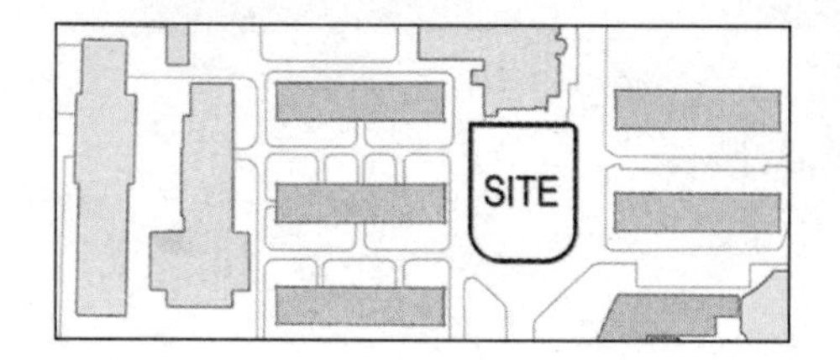

The site is located between several dormitory buildings and a canteen on the north.

We aimed to pay more attention to this kind of people. They faced with chaos, violence, alienation and irredeemable loss in life, felt propelled to leave physical traces and markings on their bodies, in order to preserve and corroborate a pure and sensitive mind from within.

Imagery of Human's Existence

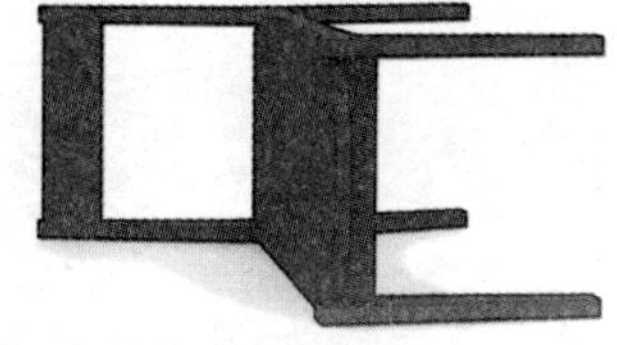

Non-Human Scale Chairs =Those Who Act Differently

There are several different relationships between this kind of people and the normal people. Some of them are defeated by the pressure of the society, some of them dissociate from the people around and some want to fit into the world but never succeed.

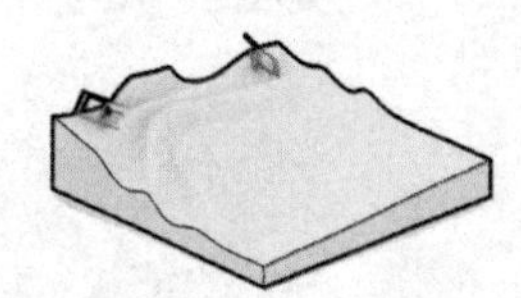

Defeated

Dissociate

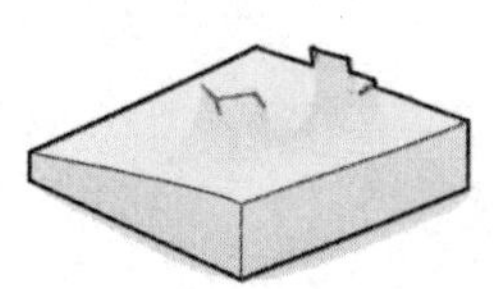

Never Fit

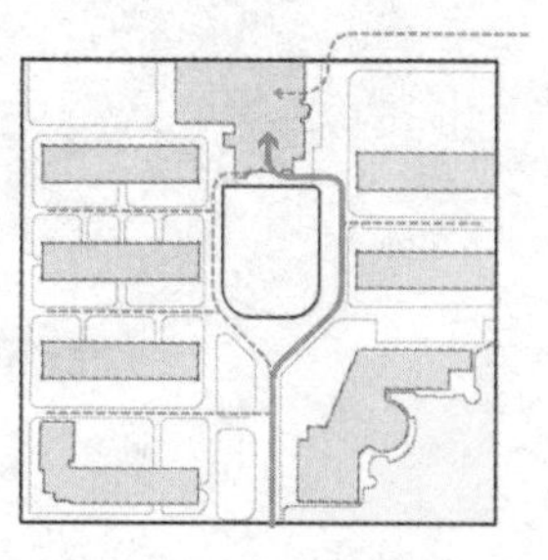

11:30-12:30 & 17:30-18:30

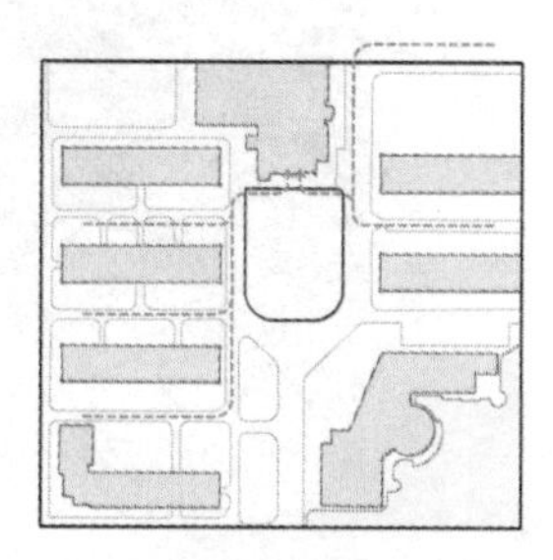

Other Time Periods

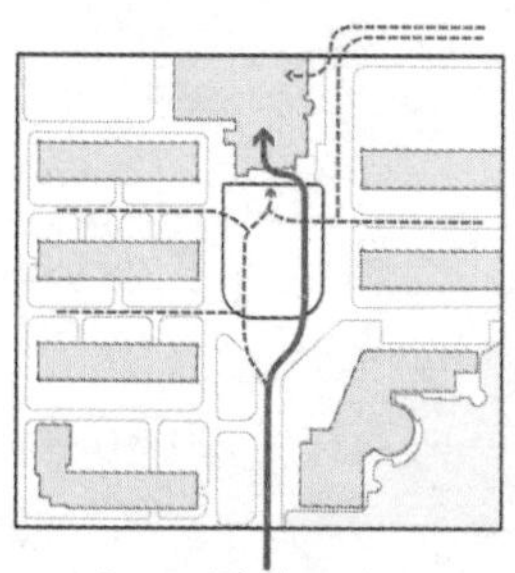

Potential Flow Lines

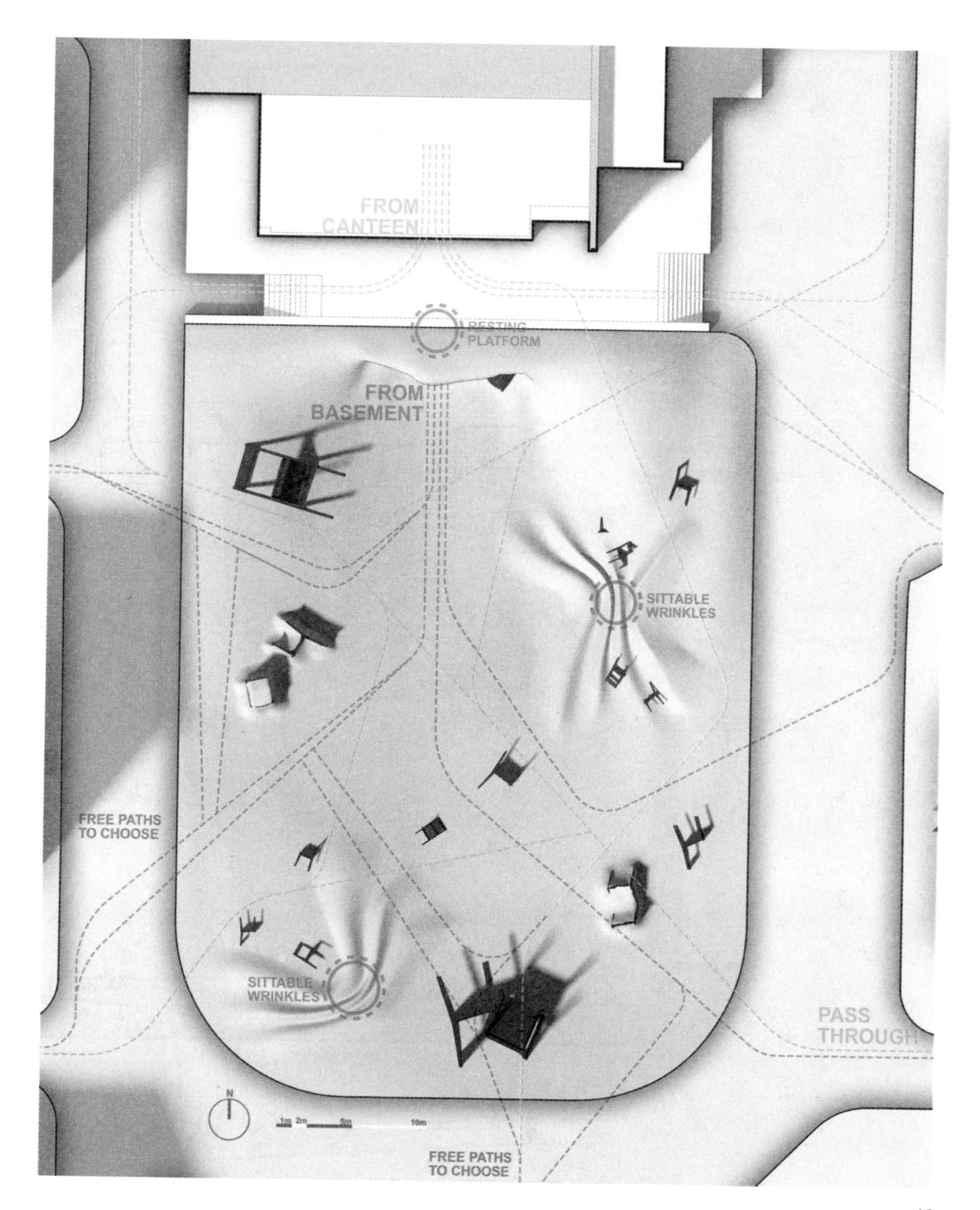

This is the project plan which shows the function arrangement and possible paths.

This diagram demonstrates the formation of the landscape. This project connects the activities in different level and makes the site more accessible.

A variety of activities can be done because of the integration of the land. Different sizes of chairs provide more functions. And one can choose the way wherever he wants to go.

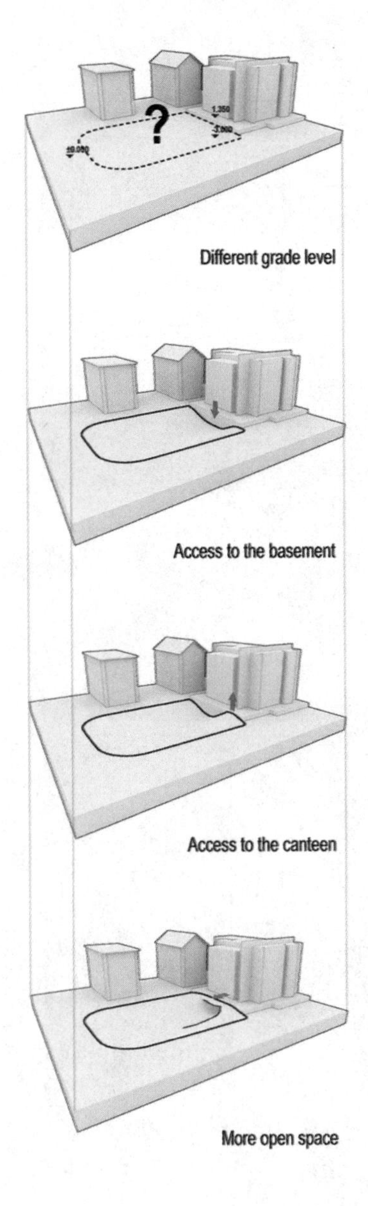

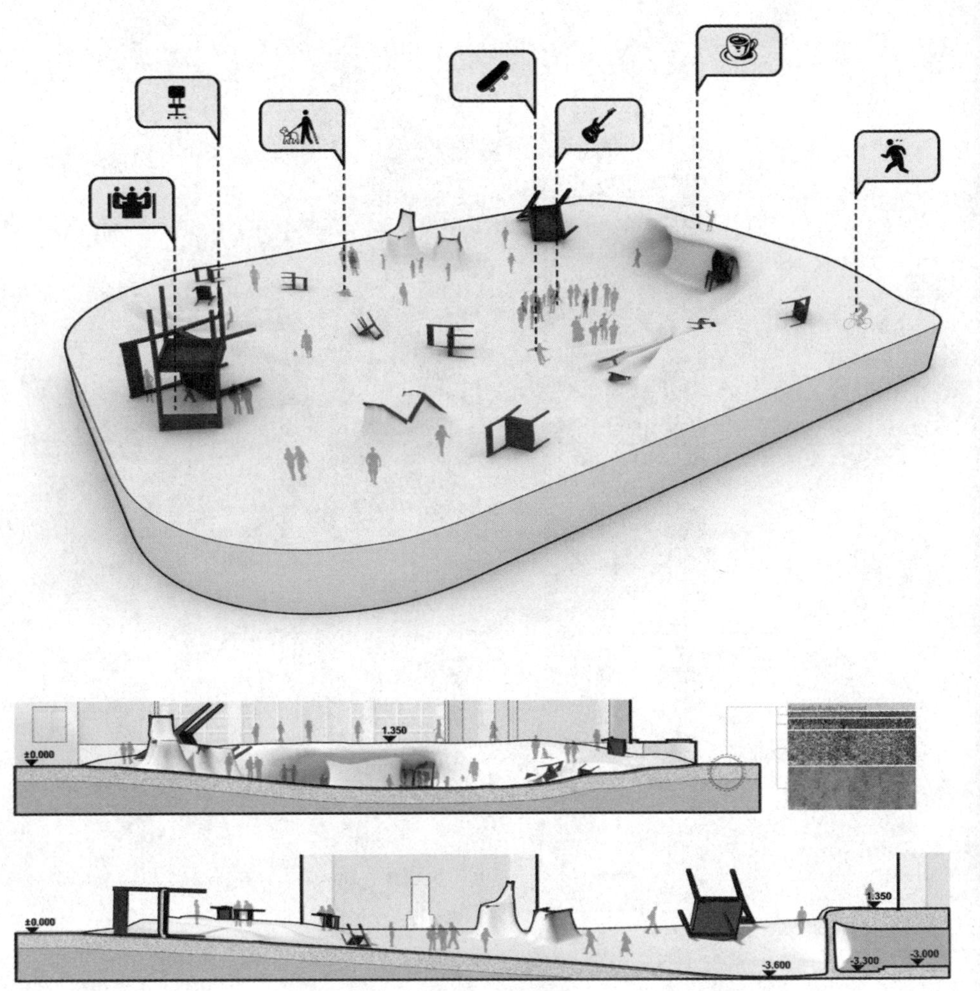

Bird's-eye view of the site shows a good description of the scene.

Several photographs of the model show the different kinds of activities which are available on the site.

The theme of the project may involve in some negative emotions. But the plaza doesn't need everyone who use it to pay too much attention to what it wants to express. The most important function of it is to provide a wonderful place for people.

I'M FINE

南晶晶、王凤阳、张薇

Nan Jing jing, Wang Fengyang, Zhang Wei

Yang Xinguang who was born in 1980 in Hunan province, lives and works in Beijing.

His works primarily use wood, stone, metal, and other materials to emphasize the connection between human and material forms, and to give visual form of social anxieties. In his work "Hello", he anthropomorphizes two wooden rods, retaining traces of having undergone manual, labor-intensive processes and symbolizing impoverishment, to hint at some of the unfair relations behind the process of globalization.

Inspiration Source
Our source of inspiration is the work "Hello" which uses abstract symbols to reflect the unfairness behind the process of globalization, then we think about the modern man, they all have all kinds of pressure, such as employment, study, family and so on. Most of the time, we have personal reasons, however, we try to keep smiling and say: "I'm fine."

Process of Creation
Our design abstracts people of this state. We extract the"smile"and"I'm fine"two elements to express contradictory emotions.

The first two letters are higher than the last ones. But in the plan view, the first two letters are smaller than last ones. It represents people are entangled in this situation. They are struggling.

In the facade, the change from high to low represents the emotion ups and downs, finally it is covered and suppressed by many smile, reflecting that our inner world is not very happy, is full of contradiction.

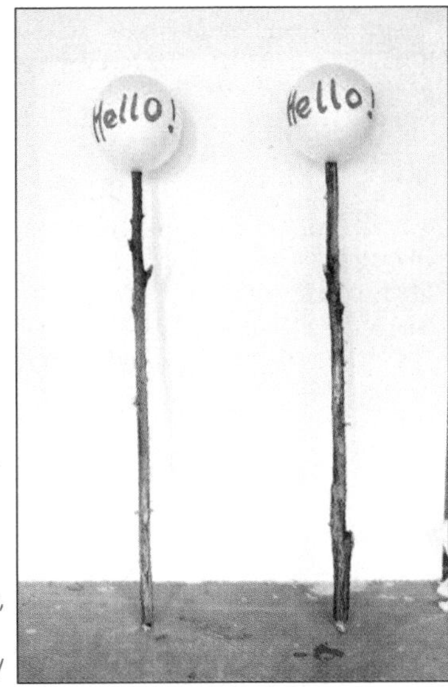

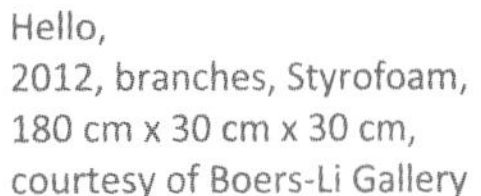
Hello,
2012, branches, Styrofoam,
180 cm x 30 cm x 30 cm,
courtesy of Boers-Li Gallery

Design Concept

The design brings users weird and ironic feelings by contracting the quantity and size of the smiling lips and "I'm fine". The lips are also used to lead the route. They converge slowly and become bigger and bigger approching the huge lip stage. Uplift of the terrain can be regarded as the auditorium.

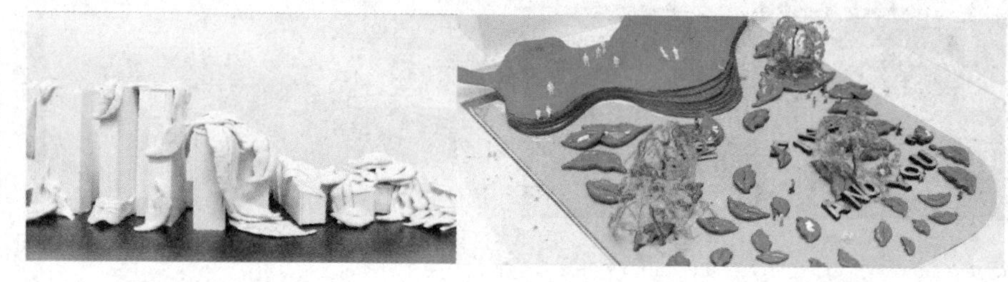

Layer Analysis

Through the art piece we choose, we feel anxious, ironic and creepy. We put these emotions into our work. The undulating pavement represents the happiness and misery in our life, hints at the trace of labor.

People always say "I'm fine" to comfort themselves. Maybe they want others not to worry about them or just because of the privacy or vanity.

The smiling lip is an irony of the status in life. We hope it can make people rethink and focus.

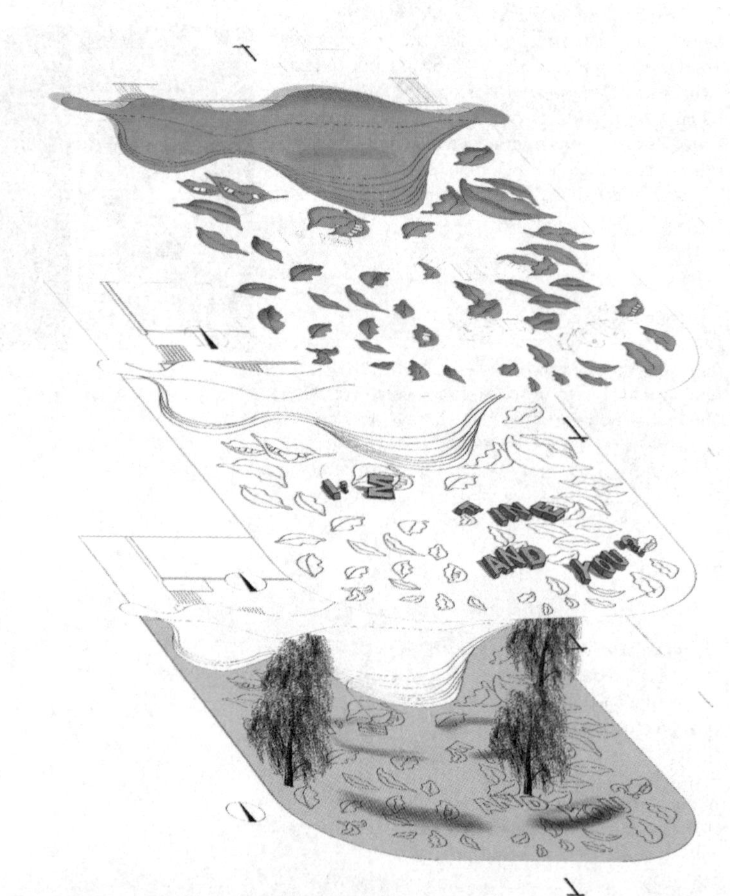

Current Problems

1. The stage lacks function;
2. The sunken plaza lacks attraction;
3. The main entrance of the canteen and plaza are apart.

Solving Strategies

1. Natural Stones stage can be used as a dining area when there is no preformance;
2. Same level with the horizon green gentle slope lead the traffic lines to free and diverse functions;
3. The main entrance of the square enter directly.

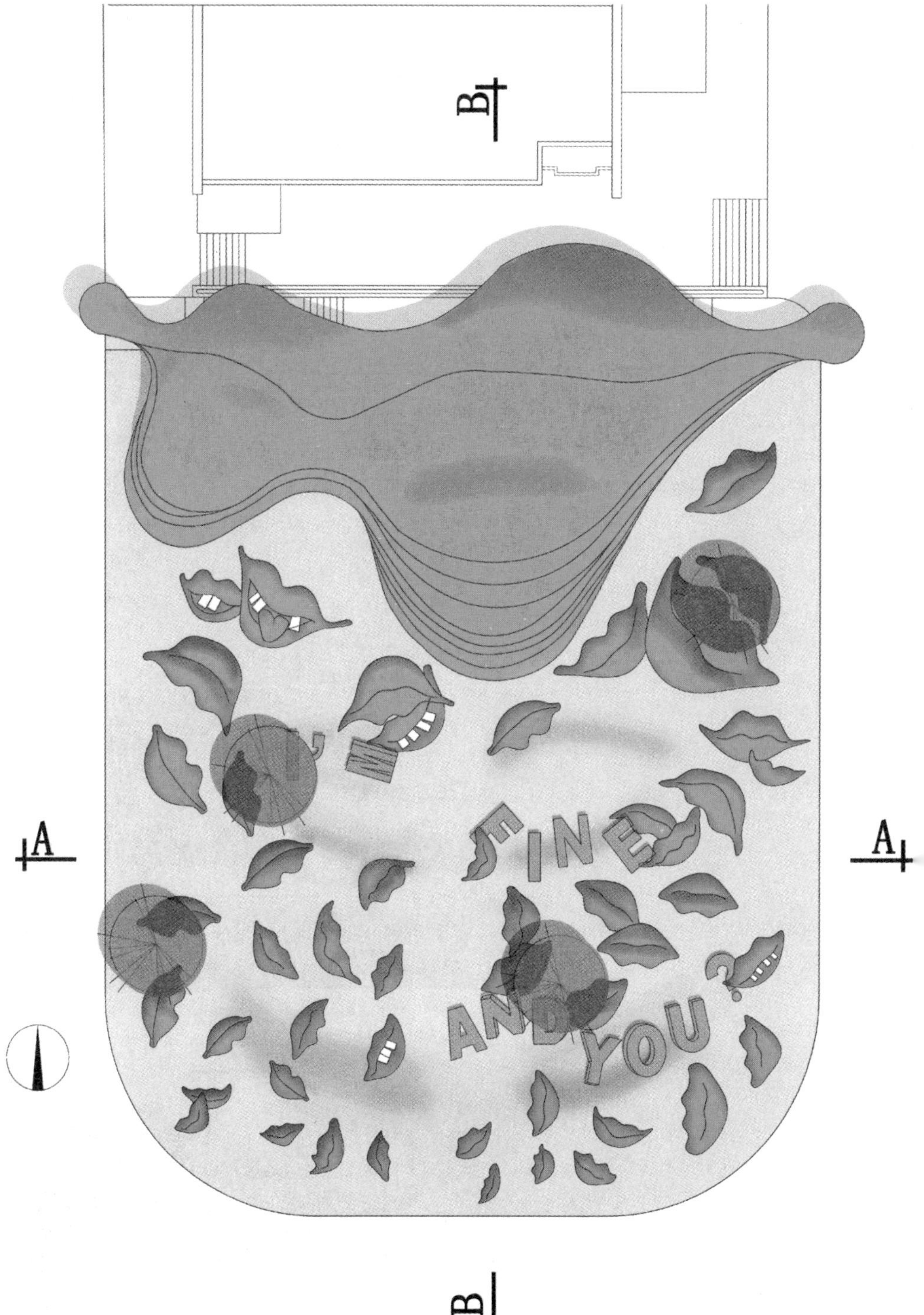

Plan

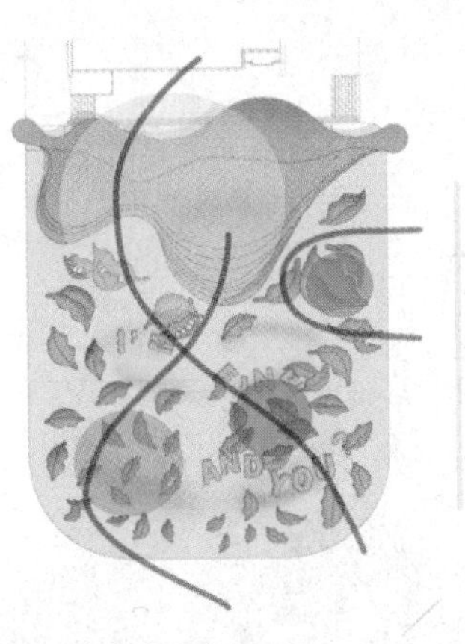

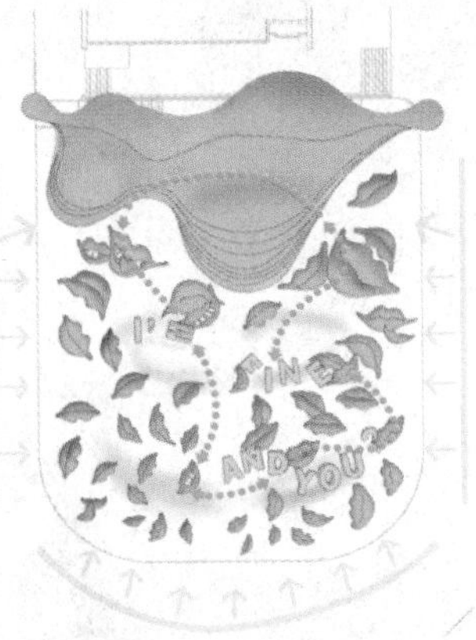

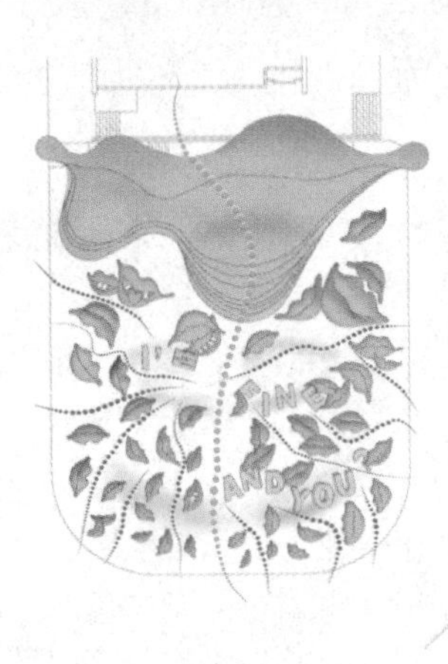

Functional Analysis

SPORT AREA
RECREATION AREA
STAGE
PLATFORM
FUNCTION CONNECTION

ROUTE VISION
VISION CONNECTION

MAIN ROUTE
SECOND ROUTE

Section A-A

Section B-B

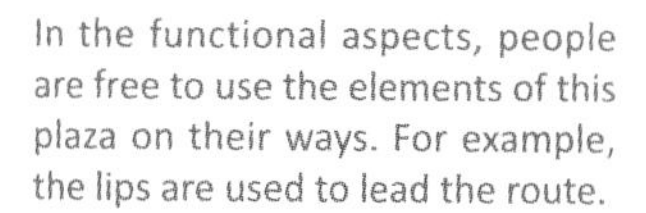

In the functional aspects, people are free to use the elements of this plaza on their ways. For example, the lips are used to lead the route.

People can use the huge lip as a stage and when there is no performance, it's a dining area.

Move the small lips to talk close, do sports and play games.

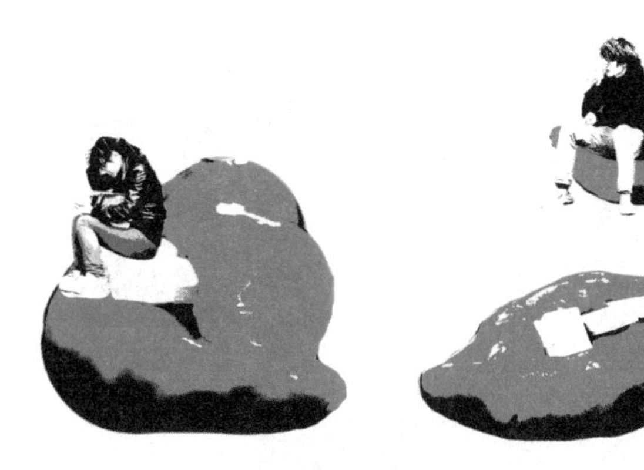

Or just have a rest, like sunbathing on the lips.

The view from the entrance to see the direction of the canteen entrance.

The lips converge slowly and become bigger and bigger approching the stage.It makes people feel that the plaza is more far-reaching than the actual situation.

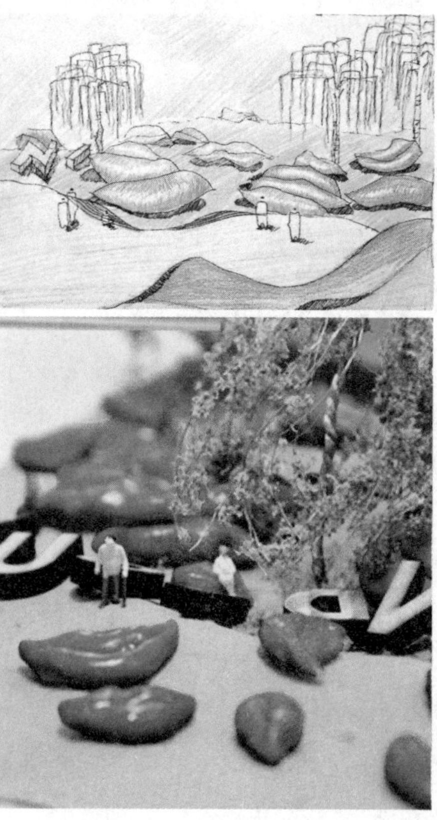

REVOLVING STRIP

郭菲、刘妍、刘燕
Guo Fei, Liu Yan, Liu Yan

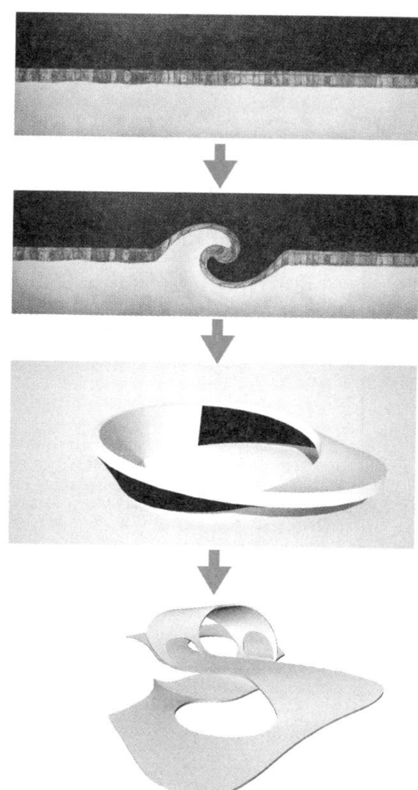

Therefore, in the process of our reshaping design model, we try to put the whole space into an infinite space. It reminds us of "Möbius Strip" which is interacting and dividing the upper space and lower space closely. This strip is featured by traffic route which has no start, no destination. We use this strip form to act as the fences to reshape a vivid and fluctuant landform with consistent traffic line and 3-Dimensional spaces from the upper canteen building to the lower basement and shop. The middle plaza becomes a partly subsidence plaza, attracting students dirrectly to the store. Eventually, they will find mysteries at the back of "those fences" (the strip).

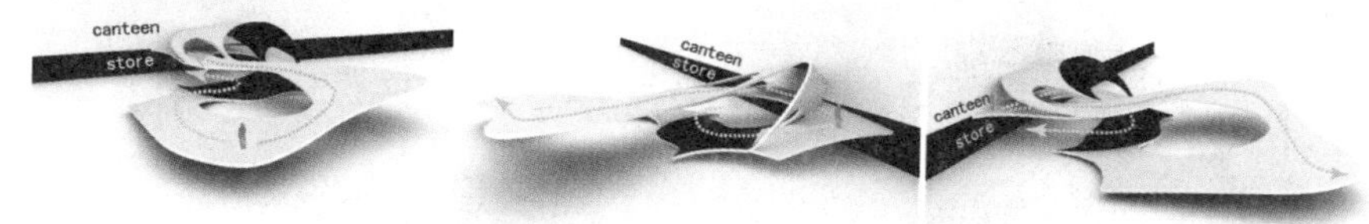

From the three perspectives, we can see respectively how this spot could be orderly joined to the other two space — canteen and store.

From the original art piece, the fences in the middle connect the upper space and the lower space, in the meantime, separate the two spaces. However, no one would actually know what it is like behind those fences, perhaps it is dark black, light white or an infinite world. Just like Forest Gump once said:" life is like a box of chocolate, nobody knows what you gonna get." You never know what it gonna happen to you tomorrow, there are always limitless opportunities and challenges.

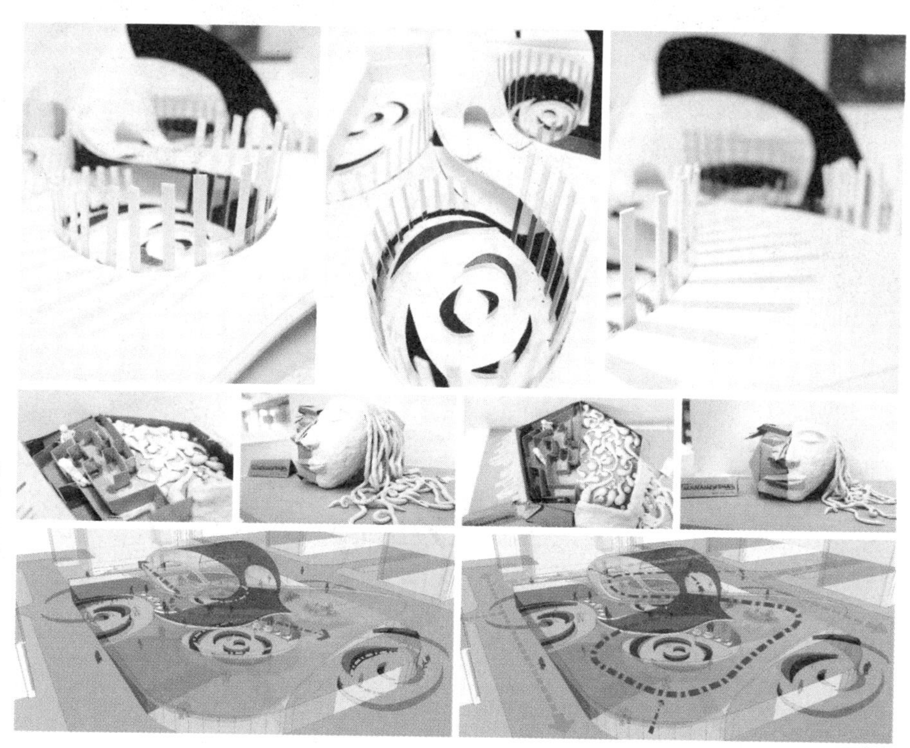

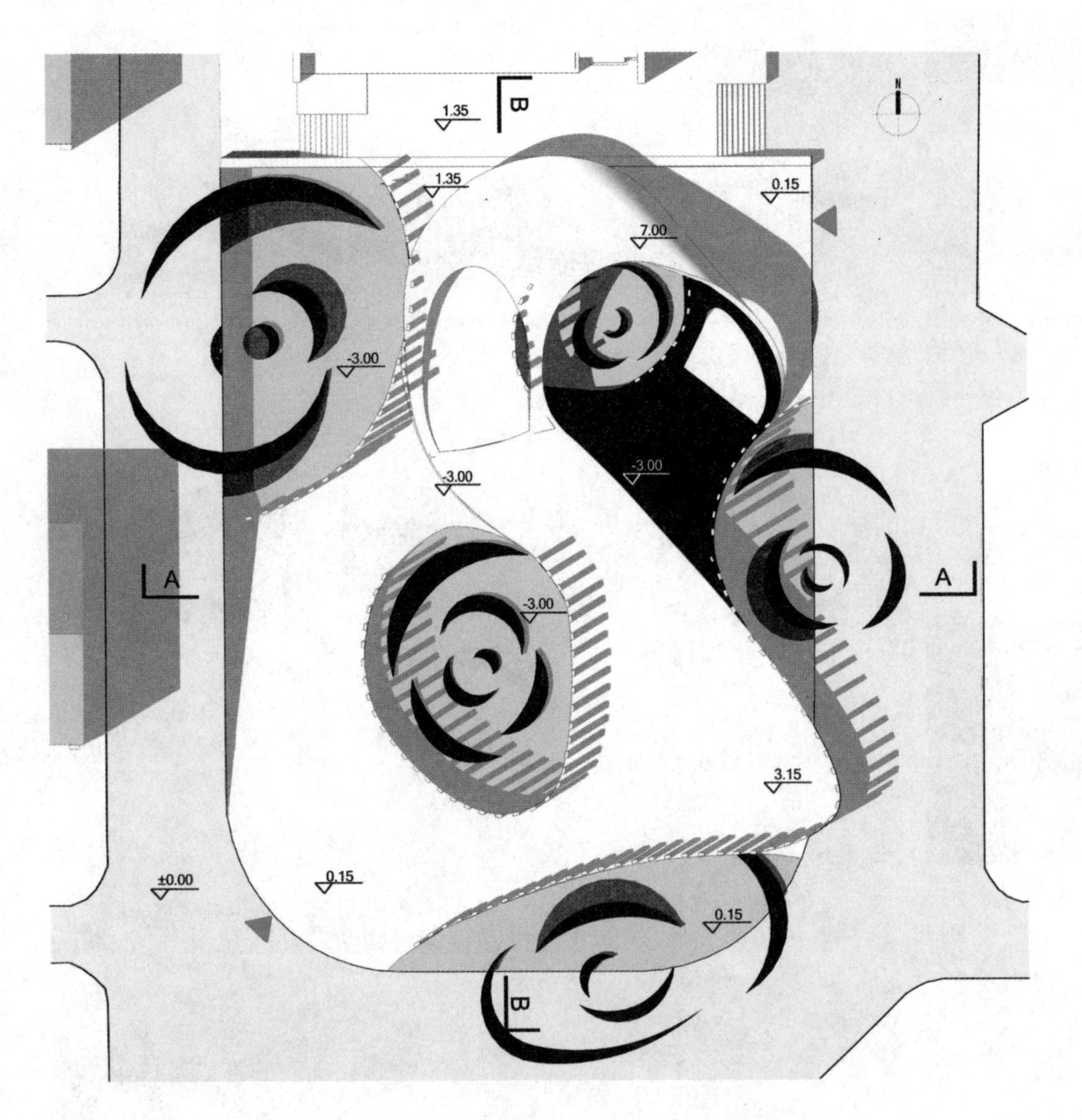

Site Plan

Functional Analysis

Activity Content Planning

The activities from the upper space and subsidence plaza are as follows: run, dance, picnic, hiphop, bicycle and so on.

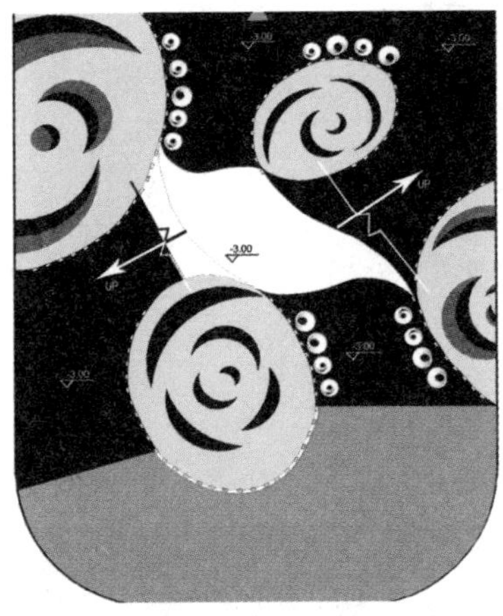

Underground Plan

Impression Drawing

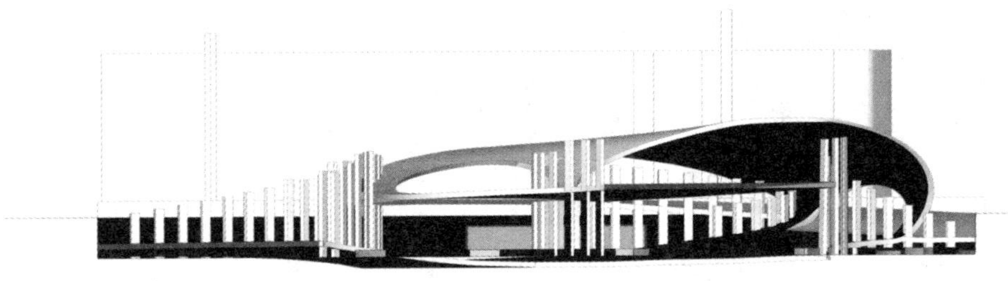

Section A-A

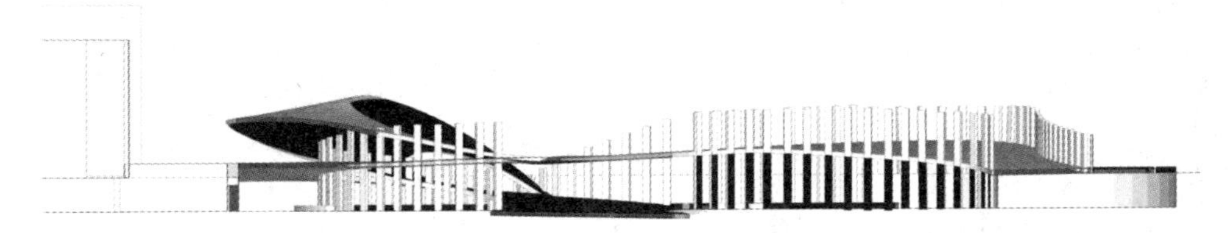

Section B-B

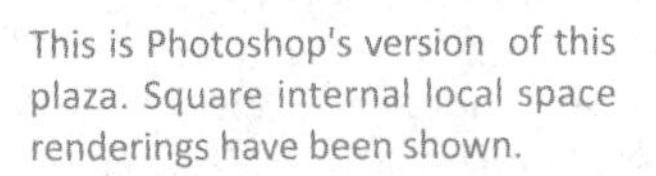

This is Photoshop's version of this plaza. Square internal local space renderings have been shown.

The original art piece, the fences connect the upper and lower spaces.

教师论文

盛强
张春彦
何捷
张昕楠
张秦英
胡一可

物体的去叙事性与去功能化——感受玛莎·施瓦茨景观设计工作营

盛强

“景观可以是任何东西！”（Landscape can be anything!）

在整个工作营结束后很久我才知道这竟然是玛莎·施瓦茨这个工作营的正式名称，曾经设想过诸如“当代艺术与景观设计”之类的名称瞬间就觉得太过于正式、古板了，完全不“玛莎”。要想贴切地用一些惯用的术语来评价或描述玛莎·施瓦茨的设计和授课方式并不容易。现代艺术也许是该工作营的一个切入点，但她对于艺术的关注绝不在于风格和形式语言的借鉴，而更多的则是帮助学生了解一种开放式的，但又不失逻辑性的设计思维方式。从这个角度上讲，这次联合教学过程中我所感悟到的东西可以更直接地应用于我之前对专题设计“符码转译”的探索中去。当然，撇开一切的功利，从参观798的作品，到整个工作营过程中对艺术品所表达的理念及其手法的讨论，这一切仿佛又把我带回到了大学时代。经历了硕士、博士生活后，被各种理性量化的模型和统计淹没之后，重新回到曾经为之疯狂的现代艺术中去是一种久违的体验。但有趣的是，也正是由于阅历的丰富，现在回过头来再看这些装置、构成和影像反而觉得比当年清晰很多。艺术与科学也许本来就是相同的东西：都注重概念的简明和逻辑的清晰，也都期待着能给观者带来不同的视觉感受，甚至惊喜或震撼，差别也许仅仅在于表达的语汇不同。

“回到你选的作品中去，思考它何以能打动你”大概是这个工作营中前半部分玛莎·施瓦茨最常说的评语。而后半部分，在涉及具体的景观设计方案后，给我印象最深的是关于建筑或者说景观设计语汇与叙事性差别的讨论。“讲故事”是大部分学生在一开始介绍概念时常用的方式，但建筑、景观，也包括视觉艺术品对概念的表达方式与讲故事有明显的不同。在这一点上现代艺术表现得可能更为直接：没有设计者可以站在自己的作品前长篇大论地向观者讲述他作品的概念，而概念必须通过作品的材料、特性、组合等能被观者直接感受到的东西进行传达。因此，“回到你选的作品中去”的目的不是追问作者的初衷，而是作为观者的感受。不同的形式、材料、颜色都可以传达给观者特定的感受，而一个能够打动人的作品往往是围绕着核心体验和概念的不同物体的创造性组合。玛莎·施瓦茨在这个工作营中希望训练的应该也正是这种对概念的专注和对物体自身表现力的运用技巧。

对物体自身表现力的强调消解了功能对形式的统治地位。“一件不能被直接使用的物体并非没有功能，被看、被感受也是它的功能。”这个观点与王澍当年做“八间不能住的房子”时体现出对功能主义的反感如出一辙。整个工作营学生被迫放弃了惯用的从功能、流线和行为分析来切入设计的方式，而完全从一件与基地不相关的，但能打动内心的艺术品出发来开启设计。设计过程中也不强调功能理性对形式的影响，更多的集中在对感受的传达是否直接、有效的讨论上。当然，这种设计方法本身是否合理是见仁见智的，但作为一种特定设计方法的探索和技能的训练，这个工作营在我经历过的各种类似的联合教学实践中无疑是非常成功的。

“别处”的风景与艺术
——艺术创作与景观场地认知

张春彦

一如“生活在别处”[出自法国诗人亚瑟·兰波(Arthur Rimbaud)(1854—1891)的诗句]，从某种意义上来说，风景园林所创造的景观作品也是一种“在别处”的艺术，即使历史上最早的园林，也是尽可能模拟“仙境”抑或“画境”，都是梦想中的“别处”。从美学的角度看，真正的生活应当永远在别处。当生活在彼处时，是梦、是艺术、是诗，而当彼处一旦成为此处，崇高感随即便成为生活的另一面：平淡无奇。所以人们总是喜欢创造、游览、欣赏别处的风景。身在此处时，我们对身边所熟知的景物，由于过于熟悉，往往会视而不见、习以为常。我们的感官会变得麻木，而麻木的人是不会欣赏和感受到风景的美的。所以作为景观设计者，我们必须尝试转换视角，重新来审视我们的生活环境，以艺术的手法，将位于此处的人们“转移”到彼处。

当人们讨论艺术，尤其是现代艺术与景观的时候，更多的是从色彩、形式等方面加以阐述，诚然现代的景观创作从现代艺术中汲取了诸多的灵感，丰富了现代景观艺术创作的手法。然而一方面景观不应仅仅是色彩与形式的表达，它应该是叙事性的，要将人置于场地之中，使人与场地内各元素以及各元素自身之间都产生对话，进而将时间与空间有机结合，“生”出许多“故事”来。另一方面，更为重要的是怎样才能对我们熟视无睹的场地产生兴趣，从而发掘、发现场地的特征呢？设计师一己之力似乎已经无法解决这个问题，于是人们经常提到公众的参与。而现实是，生活在此地的人们往往如同我们面对的场地一样沉默，读不懂，也体会不到什么，他们内心深处都是“别处”的风景。在还没有真正认识场地之前，作为设计师我们是拷贝一个别处的风景过来，还是什么都不做，又或者我们还有第三种选择？

我们来看一个案例，法国西部大西洋沿岸卢瓦尔河串联起来的两座城市——南特(Nantes)与圣纳泽尔(Saint Nazaire)(图1)。南特是近现代法国西部重要的造船基地，其造船业经历了发展、兴盛到衰落的百年沧桑。随着城市的港口功能逐渐被位于卢瓦尔河入海口的圣纳泽尔市所代替，城市只留下少量的内河航运功能。如何衔接两座城市？它们之间的纽带——卢瓦尔河随着航运的消失已被人们“视而不见”，尤其是不能武断地抛开公众的意见而创造一种新的功能，生硬地将二者连接起来。现代景观艺术作品的创作给了我们第三种答案。

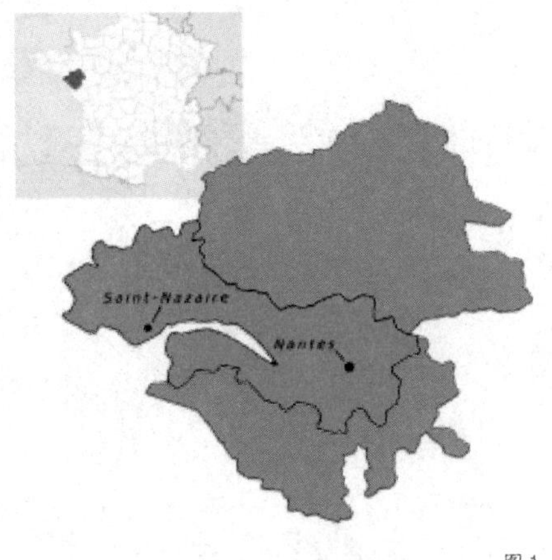

图1

在南特城市复兴计划中制定了卢瓦尔河沿岸风景园林规划，该规划邀请世界各国艺术家们来进行风景园林艺术作品的创作，这些作品线性地分布于南特及圣纳泽尔城市内的文化场所以及沿卢瓦尔河的两岸（图2）。作品与周边的风景一同吸引着公众的视线，在参观、欣赏这些艺术作品的同时，公众不由自主地将自己置身于这一方土地之上，不由自主地参与进了地域的演变，变成了过程中的一部分。仰视、俯察、远望、近观之间逐渐思考、领会地域的尺度及复杂多样性。其中某些作品将永久地保留下来，这些艺术作品极大地激发了地方文化活力，冲破了人们对自己生活地区的“惯性”思维，从而能够带来两座城市衔接的多种可能。

所以，现代风景园林艺术家们触及一个场地的同时，借助于有形及象征性的背景，于此处建立起彼处的意象，他们改变了我们与场地及其建筑景观之间的关系，他们的作品是另外一种模式的感受与再现，另外一种模式对地域场所的解读，提供了另外一种视角来解读、认知此处的场地，将我们从此处转移到了彼处，使我们能够更好地认清此处场地的价值所在。

参考文献：
张春彦，乔羽．风景园林学科在法国城市发展建设中的作用［J］．中国园林，2012，5.

图2：卢瓦尔河沿河景观艺术品

来源：ESTUAIRE, NANTES SAINT-NAZAIRE，2007

现象学视野下的景观作品创作与认知

何捷

回顾本领域中被广泛认同的景观定义，绝大部分都强调了人与自然的互动，典型的如《欧洲景观公约》，强调景观是为人类所认知而产生的现象学意义。在本次玛莎·施瓦茨教授的设计工作营所倡导的基于艺术主题的，自直接、先验本质转译的景观环境与空间，正是现象学在景观生成和景观欣赏这一框架下的典型表现。

然而设计师所依据的先验或分析的生成逻辑和结果，实际上只是部分的景观建构。景观作品建成后被使用者和公众的认知和欣赏——或者在本书所展示的设计工作营的整个过程中，被评图教师甚至一同聆听方案汇报的同学，乃至本书的读者的解读——在这一景观作品整体意义上是对其余部分的完型。在这种语境下，景观作品从设计到建成和使用实际上是一个在景观意义上不断变化的过程，设计师先验的设计概念究竟在多大程度上代表着景观作品的价值核心，是值得讨论的问题。

如果进一步将景观作品置于历史维度中，当设计师的原本概念湮灭、抑或被记录下来而成为景观历史文本中固有的一部分，观者的认知则成为这类景观建构的主体甚至全部。从景观设计的生成、作品的营建完成到更长时段的存在，建构作品整体意义的先验概念和后天解释之间一直在保持此消彼长的动态平衡。

针对自然景观的视觉景观评估方法论中，“经验学派”即是基于现象学理论和认知模式理解评价偏好，但其在保证高可靠性与客观性的同时带来具体操作上的困难，实用性远远不及最“不科学”的“专家法”。崇尚严谨的科学研究尚且如此，景观设计与设计作品的评价则可能更为动态。究竟设计师和不同受众针对同一设计作品和景观对象认知和解读是否具备某种程度上的一致性，或者说作为景观作品的同一物质环境和物质现象的心理映射过程的规律性，对于更具内涵意义的、创作的景观设计作品来说，显然是比自然景观的规律性偏好更加复杂的问题。

回溯玛莎·施瓦茨教授此次设计课所强调的艺术品至景观作品的生成线索，其设计生成过程本身就是经验主义的，每个步骤都可以认为是一项景观作品。因此对作品生成逻辑的建构，也可以认为是学生和指导教师合作的结果。由于书籍作为这些作品的主要文本形式，设计者的先验观念和设计生成的单向解读成为建构这些纸面景观作品的主体。然而随着本书的面世，读者对这些作品的理解与评判，乃至将来学生们其他景观创作作品的实现，受众的认知将增加景观作品意义的复杂性。在景观设计与生成过程中，了解未来景观重新建构潜力和动态性，可能会对设计方案和设计技巧本身有不同的理解。

一场有关艺术的游戏
——记玛莎·施瓦茨景观设计工作营

张昕楠

在当代著名景观设计师中，玛莎·施瓦茨无疑是最特别的一个。她的作品充满了独特的艺术气息，而且往往以一种游戏化的处理方式带给人们充满童趣的深刻体验。她的每一个设计，在以装置对环境做出景观梳理提供艺术感受的同时，也为人们提供了一个有趣的“游乐场”。

此次玛莎·施瓦茨景观设计工作营一如玛莎·施瓦茨的作品，为学生准备的也是一场有关艺术的设计游戏。在教学过程中，她引导学生对艺术进行深入解读，以形式创造对艺术的感受进行复述、再生，以景观设计永载艺术的灵魂；这样的教学方法，训练了学生对艺术的敏感把握、对直观感受的形式表达，使学生在“耳濡目染”于艺术的过程中提高创造空间环境景观的能力。

在希腊语中，游戏(paidia)和教育(paddies)的词根一样，都是指儿童 (pais) 的活动；杜威也认为“教学的第一要素是给学生一个真正感到兴趣的地方，没有一些游戏，就不可能有正常、有效的学习”。玛莎·施瓦茨的教学，交给学生的不是枯燥的设计任务书，而是一件他 / 她自己挑选并喜爱的艺术“玩具”——如此一来，教师在教学中完成的其实是引导学生如何“玩耍”的过程。更为重要的是，这一环环相扣的“玩耍”过程，有效地训练了学生的艺术、感性认知，并将这种认知潜移默化地融入学生的设计训练中去；这同传统教学强调以固定课题训练学生的设计手法相比，更强调学生的自我主体性，在教 - 学的关系中更加向学的一端倾斜，着重引导学生以某一感受为出发点掌握设计的方法，以艺术为媒介有效地激发学生的学习和创作热情。

虽然在历史上，东方以诗、书、画、意的艺术结合完成了园林的文化精粹，西方以科学的美学法则、艺术的画与像萃取出环境自有的场所精神。然而，当代以来，随着艺术形式、定义的颠覆和多样发展，以及信息化社会读图化、快餐化的文化倾向，越来越多的设计或以“语不惊人死不休”的形式夸张地完成着“巧言令色的视觉屠戮”，或以“因为 + 所以 = 高效”的理性霸权实现着机械的全球化，悦目的艺术创造和赏心的敏感思考似乎与设计渐行渐远。在这一背景下，本次工作营为我们提供了一个重新审视景观设计本质和景观设计教学的机会——我们应如何在理性 - 感性、功能 - 艺术的博弈中再一次寻求内心的平衡?

正如塔夫里曾经说过的，“功能的实用生于理性，艺术的冗余发自内心”。

生态与美——关于景观功能的思考

张秦英

三周的时间，我基本全程参与了这次的设计课，毕竟机会难得。同之前所了解的玛莎·施瓦茨一样，她专注于从艺术的角度寻找灵感、提炼主题，甚至于最后的设计形式及景观功能的实现。不同的是，这次近距离的接触、交流，扩展了我对艺术的理解。

“美，本身就是一种功能。”

可能与本身自然科学的知识背景有关，我在评价一个设计方案的时候，会过多地关注其使用功能。当玛莎·施瓦茨提到“美感就是一种功能的实现”的时候，我受到了触动。然而对美之功能的实现，需要设计者及使用者更高的艺术素养，这需要景观教育及全民教育共同努力才能实现。

“我在每个作品中都是充分考虑生态性的。”

对她以往作品的评论，更多的是过于追求艺术表现而忽略了生态功能。我们在此次活动的研讨会上也提到了这个问题，对此，她也表示无奈，她说道：“在我的每一个作品中，从理念到材料的选择，无不将生态性考虑到其中。”进一步分析，在她的很多作品中都使用假的植物而非可以固碳的植物材料，这可能是评价者认为不生态的主要原因。低碳、降低碳排放、固碳等与CO_2相关的词汇成了生态的代名词，植物固碳释氧的作用又使大家认为栽种植物便为生态。于是乎，使用假的植物就变成了不生态。对此，她解释到：可能是因为我面对的项目，多是在不可能做景观的地方，甲方低投入、零养护的要求，使得我不得不这样选择。撇开甲方的要求，是不是在任何做景观的地方都需要用到真的植物呢？种植植物一定就是低碳和生态吗？比如她的作品之一怀特海德研究所的“分裂园”(Whitehead Institute Splice Garden)，这个屋顶花园不见阳光、没有水源、没有养护人员、低预算，所有的植物都是塑料的。如果我们非坚持用活的植物材料，便要提供后期养护的设施，给水排水、修剪、病虫害防治，以及后期涉及死亡植物材料的更换。如果从碳的角度计算的话，真植物的使用不是固碳而是排碳了。也许这是一个极端的案例，但可以折射出我们处理景观植物种植设计的问题。在高层住宅屋顶建设复杂的花园、城市中大量的大树移栽等，这些植栽项目，貌似为小环境提供了固碳的材料，然而使植物正常生长付出的碳排放远远高于其本身的固碳量。

植物本无错，错在营造者没有认真地做好数学题。

所以，通过这次设计课，从玛莎·施瓦茨身上，不仅学到了如何看待艺术，也体会到了生态的艺术。

别样的世界

胡一可

景观设计为人服务，而当今景观设计的对象多为无特征的标准人，对于风景园林师而言，其性别差异对设计作品的影响也较为明显。玛莎·施瓦茨教授的作品流露出显而易见的女性特征，其根源在于女性的生理特征、心理特征、审美心理，让参与此次教学活动的师生在关注诗样体验、情感传达、场景再现和互动参与方面受益匪浅。

当代的景观设计面对的是平均身高、有正常习惯、身体健康、异性恋、无自身特色的标准人。如果说设计是为人服务的，那么上述现状是造成景观设计品质粗糙的重要原因。对于使用者如此，对于设计师本身亦如此，不同的设计者会创作出不同的设计作品。毕竟，每一个单独的个体都有其独特的生活经历，从而导致了其不同于他人的价值与审美评判标准。男性与女性是人类众多差异中最明显的一个，而性别差异也使设计作品产生了一定的差异。由于男性占据社会主导地位，女性价值和女性特征往往被忽略。从社会性别结构来看，男性与女性是对立而又互补的。在设计领域，现阶段很少关注性别差异，可以说是一种无性别差异的设计。本文关注女性意识和女性设计师的价值取向，基于此对女性风景园林师的作品进行分析，并总结出相应特征。[1]当然，这些特征在男性设计师的作品中可能也会出现，本文并未纠结于此，而是从新的视角解读女性风景园林师的作品，为风景园林设计品质的提升提供有益的参考。

一般来说，女性的触觉、味觉、视觉和听觉反应大都比男性敏感，其原因在于女性特殊的内分泌构造[2]；调查问卷显示，女性更能感受到空间的特质[3]；女性接受暗示的能力较强，因而在没有外因作用的情况下比男性更易接受新的习惯和观念；女性大多不擅长理性分析，对抽象的理论缺乏兴趣[4]；由于能够及时地利用情感复合体进行宣泄，女性不容易受到严重病症侵害。基于此，具有女性特征的设计往往体现出细腻、缺乏理性、空间特征明显、具有趣味性等诸多特征。

由于女性承担了繁衍后代、延续人类生命发展的重要使命，她们的感受性先天与自然界的发展趋于统一[5]，在探讨景观设计与自然的关系过程中更具优势。一些女权主义者认为女性的从属地位是因为女性在生理、心理和社会角色上更接近自然。而自然万物是艺术创作的主体，女性的创作很多包含对大地的情感。女性亲近自然，注重感性的、细微的变化，她们对景观要素的形体、色彩、光线、质感等都有着独特的感受，因而也更容易受到环境的影响。

以玛莎·施瓦茨为代表的女性风景园林师的设计作品特征解析

诗样体验

女性具有较男性更为复杂的神经系统，因此对空间的情感特质十分敏感，有利于其营造空间氛围；细腻入微的情思给女性带来更为丰富的情感世界，同时也会产生某些易变、非理性的设计。优秀的女性风景园林师会对景观进行感悟式、体验式的书写，并用独特的叙事视角、意象化的方式和鲜活的设计语言加以表现。

诗样的体验还意味着将时间概念引入作品，并作为设计的关键因素。当观者接近作品时，可直接感受到作品的内涵，这种内涵因人而异，而空间在时间中流动，形成风格别致的诗性美学景观。

互动参与

女性较之男性有着更强的好奇心，更喜欢参与，女性风景园林师的设计关注景观本体与观众的对话。这些作品依靠于本体的或通感的，而非后天的反应，来让观者体验，作品重触觉质量。观者深入景观作品时，某些行为将自动发生。

在当代景观设计师中，玛莎·施瓦茨无疑是特别的一个。她的作品充满了独特的艺术气息，而且往往以一种游戏化的处理方式带给人充满童趣的体验。她的梅萨艺术中心设计，在对环境进行景观梳理并提供艺术感受的同时，也为人们提供了一个有趣的“游乐场”，项目建成后她经常乐此不疲地欣赏使用者体验这一景观空间（图 1）。

情感传达

女性独有的特征使其设计作品成为情感传递的媒介，女性设计师多偏爱独有的空间语言，注重在公共场所中营造私密感，观者的阅读行为也变得更具亲和力。女性设计师的想法和意图更易于直接传达给另一个人，不需要进行翻译。

在成长过程中女性的心理和潜意识不断变化，女性特有的敏感心灵能更真切地体验灵魂的迷幻、变异现象等。具有女性特征的景观设计历程也是女性心灵史的书写，基于女性独有的性别经验和丰富的情感世界之上。玛莎·施瓦茨在英国伦敦设计的圣玛丽亚教堂庭院（St. Mary's Churchyard）是在情感传达方面的极好案例。

场景再现

相对男性视角，女性视角更注重对生活细节的把握，融有更多的感性因素。优秀的女性风景园林师创造供人思考的场所，促成属于观者自己的感受，在其脑海里有着一系列经典的场景，而景观设计是一种场景再现的尝试。她们很少去支配或覆盖土地现有的景观，而是试图与现状景观互动，营造特有的场所感。

城市环境中的景观设计亦如此，正如玛莎·施瓦茨所说，城市景观与其说是一个环境，不如说是一个社会和文化的平台，提供给个人及城市一个健康的生存空间。人们需要制造场所感，确立身份认同感，玛莎·施瓦茨正是基于此，将自己童年中最美妙的场景带给世人。

图 1：梅萨艺术中心

玛莎·施瓦茨设计方法总结

形式转译

以玛莎·施瓦茨为例，她对现代艺术了解较深，其作品的许多灵感来源于大地艺术、波普艺术及其他当代艺术化表现手法。她选择用艺术的眼光对现有艺术形式进行变形和再创造，其作品是对艺术的再次解读。

她将波谱艺术观念带进了风景园林设计领域。在她的作品中，花坛、树丛、灌木、座椅等传统元素均以非常规的面貌呈现；同时，与波普艺术家偏爱使用艳丽色彩展示自己的普适性、大众性一样，她也喜欢选用艳丽色彩装点园林，她的绝大多数作品都用热烈大胆的颜色，给观者强烈的视觉冲击，比如纽约亚克博·亚维茨广场（图2）。

符号运用

玛莎·施瓦茨早期的作品有标新立异的倾向，例如波士顿的面包圈公园，遭到了众多男性设计师的批评，然而其出人意料的效果和崇尚临时感的设计理念却体现了设计师顽皮的天性，如今面包圈公园已经成为景观设计的经典之作（图3）。她将日常生活中对现成品的应用从艺术领域拓展到风景园林设计领域，大量应用彩色混凝土、轮胎、铁轨等日用品营造景观。

许多女性风景园林师善于通过对场地历史、文化的挖掘，将个人思想与情感付诸于景观形式表现上，利用符号激发观者内心的共鸣，使其嗅到艺术的气息。而符号也仅仅是一种手段，女性设计师的创作目标是建立一种持续的暗示来丰富人们在场地环境中的情感体验，通过运用生活化的景观符号与人们进行亲切对话，回归对日常生活的关注。

图2：纽约亚克博·亚维茨广场

图3：波士顿的面包圈公园

图 4：西安世界园艺博览会大师园 6 号基地

制造幻觉

很多女性设计师坚信在当代设计中，景观不应只满足于功能的需要，也应致力于表达生活与艺术、空间与形式，帮助人们认识并体验这个世界。玛莎·施瓦茨在西安世界园艺博览会的大师园创造了一组迷宫，每个走廊的尽端都有一组 3 米 高的大镜子，无限延伸走廊的长度。游人能够走进相邻的走廊，但有时会闯入周围布满镜子的小房间，自己被多次反射，难辨方向，而此时的惊慌与无助会呈现给镜子后面的其他游人（图 4）。当游人离开时也会发现自己可以通过单向镜面秘密地观察他人。迷宫里人们的活动被真实呈现，在即将结束游园之际，游人终于明白镜子在迷宫园中的奥妙。

结语

女性的成长是一个潜在的强大的真实存在，不仅仅与其自身的发育有关，与社会、文化、心理等因素的影响也密不可分。在设计世界里，女性风景园林师不应扮演修修补补的角色，而应带给我们不同的景观世界。在艺术的敏感性与判断力方面，在情感的沟通与表达方面，女性风景园林师往往更具优势。生物具有多样性，人类社会需要多样性，设计同样呼唤多样性，本文的探讨并非要表达“男性的期望≠女性的期望”，而是提供新的视角审视景观设计，以期带来设计品质的提升。

参考文献

[1]（英）埃斯利 . 男与女 [M]. 北京：中国文联出版公司，1989：22-34.

[2] 都胜君 . 建筑与空间的性别差异研究 [J]. 山东建筑工程学院学报，2005，3（20），27.

[3]（美）霍尔奈 . 女性心理学 [M]. 窦卫霖，译 . 上海：上海文艺出版社，2000：56-60.

[4]（法）西蒙娜·德·波伏娃 . 第二性 [M]. 北京：中国书籍出版社，1998：4-9.

[5] 彭向刚，袁明旭 . 论转型期弱势群体政治参与社会公正 [J]. 吉林大学社会科学报，2007（47）. 63-70.

图片出处：

图 1：http://blog.sina.com.cn/s/blog_5dcf4eba0100tpim.html

图 2：http://blog.sina.com.cn/s/blog_6b08ffc00100mdnw.html

图 3：http://www.chla.com.cn/show.aspx?id= 73059&cid=74

图 4：http://nuomixifan.blog.sohu.com/174258432.html

玛莎·施瓦茨的演讲内容

可持续的软性层面与坚持不懈的城市景观
The Softer Side of Sustainability and the Hard Working Urban Landscape

当代对可持续发展议题的讨论大部分都集中于绿色建筑，规划师和建筑师及城市领导者很少关注城市景观起到的重要作用。如果要营造可持续而健康的都市环境，就必须考虑建筑之间的城市景观。

In our sustainability discussions, most of the focus has been on green buildings. Planners, architects and city leaders have often been slow to recognize the important role that the broader urban landscape plays in making cities successful. The spaces between the buildings that constitute the urban landscape need to be considered if we are to create sustainable and healthy cities.

将我们的景观环境本身作为一种基础设施来理解，并认为它们对城市的运行和人居环境有着非常重要的作用，这种观点是一种新兴的认知范式。目前，城市的巨型化，全球范围的城市化，接近 70 亿的世界人口，以及城市的增长和再生都使得城市景观变为一种很有价值的商品。城市景观现已成为一种高端功能和活动聚集的领域——我们都知道它的重要性，发展的同时需要保护环境并减少碳排量，这在世界范围内都是最重要的议题。

The awareness of our urban landscape as a built piece of infrastructure, crucial to a city's performance and liveability, is a very new paradigm. New mega-cities, a vast global trend towards urbanization and a world population of 7 billion, plus a surge of cities that are growing and regenerating, makes the urban landscape a much more valuable commodity in a city. The urban landscape is now a territory of high function and performance—we all know how important it is that we develop so to better protect our natural environment and offset carbon emission—this is the number one issue that we all share globally.

城市景观也会在经济方面起作用。景观可以成为城市或项目的名片而使它更具竞争力。当我们的城市变得更加全球化和均质化的时候，就越发需要创造一种新的、明显的可识别性来将周边的街区和城市区别开来，依此来制造场所感，确立身份认同感。差异和独特性赋予了城市竞争力，这对新建的和改造的城市都非常重要。

But the urban landscape also functions as an economic differentiator. A landscape can brand a city or development so to create a competitive edge. As we globalize and become more homogenous, there is an increasing need to create a new or enhanced identity that differentiates neighborhoods or cities, to create a "there, there", and establish an identity. Distinctiveness and uniqueness may give a city a competitive edge, something of crucial importance to new and regenerating cities.

城市景观和经济
The Urban Landscape and the Economy

富有经验的市长现在都把受过良好教育的人作为一种资本和资源。没有受过良好教育的大众，城市经济便不会繁荣，也会缺乏竞争力。

Sophisticated mayors now consider well-educated people as "capital" and "resource". Without an educated population, the economies of these cities cannot compete and thrive.

城市的美化、绿地空间和街道空间的可达性也是吸引高知高技工人居住和工作的必要条件。因此，公共场所的绿地是在全球化市场竞争中吸引人才的一个重要工具。市长们希望他们的城市能提供多样的、可选择的环境，这正是来此发展的人所追求的。当选择性作为可持续发展的概念之一时，人们必然会选择那些吸引人的城市而放弃那些不吸引人的城市。

The beautification of a city and the accessibility to green spaces and tree-lined streets are used to entice knowledge-based workers to come to live and work in that city. Thus, the public realm landscape is one of the primary tools for attracting people to cities so they can compete in a global marketplace. Mayors want their cities to be a city of choice, as the people they wish to attract have a choice. As choice is now an operative concept in sustainability, it is a given that people will choose attractive cities over unattractive cities.

城市景观与其说是一个环境，不如说是一个社会和文化的平台，多数的环境平台不过是以保护为目的，通过法律法规的控制条文实现的。城市景观的社会和文化功能则必须被理解为一种在城市尺度上实现可持续发展的重要组成部分，而这个部分正是我所说的“可持续的软性层面”。

But the urban landscapes is most operative as a social and cultural platform, even more so than an environmental one, as most of the environmental systems are provided for and protected through planning laws and code regulations. The social and cultural functioning of the landscape must be recognized as part of what creates sustainability at a city scale—these are the components of what I call the "Softer Side of Sustainability."

这也许在认识上是一种根本性的革新，但如果我们要把城市建设成为健康而令人期待的居住环境，就必须以一种有意义的方式来积极地处理环境问题，如提高利用效率，减少不必要的基础设施和化石燃料消耗。

This is perhaps, a radical stance to take, but if we make cities healthy and desirable places to live in, then we will help in a meaningful way to positively deal with environmental issues through more efficiencies through proximity and cutting downs the need for redundant infrastructure and fossil fuels.

城市景观在社会及文化生活方面都起到了重要的作用。
The urban landscape plays a huge role in both the social and cultural life of cities.

从旧的文化中产生新文化能力，归因于人口

和世界经济重心的变化，这些变化总是在城市中发生。而公共领域的景观环境则为这些转变提供了场所。人群的聚集不是在郊区或某人的起居室中就可以实现的。

The ability for new cultures to evolve from older ones as a result of shifting demographics and world economies is done in the cities which are typically their first port of call. The public realm landscape is the "pot" in which the melting happens. The acculturation of a population cannot be done in the suburbs or in one's living room.

公共领域的景观在城市中提供“文化生活论坛”的作用现在变得更为重要了。为吸引城市居民，城市中的文化和环境健康问题现在已经成为市长桌面上的首要议题。一个城市能够在文化上提供的东西会产生巨大的吸引力，其自身便是一个新兴产业。那些曾经在博物馆和剧院中才有的活动现在在街道和公共空间中也可以看到，在这里人们可以欣赏街头表演、音乐会、装置艺术及舞蹈。公共空间里的景观已经成为文化演出的新舞台。它的开放性和共享性反映出城市本身的活力，世界上每个角落的人都可以参与其中。

The ability for the public realm landscape of a city to provide the forum for the cultural life of a city is now of utmost importance, as the cultural and environmental health of cities is at the top of a mayor's "to do" list to attract and keep people. The cultural offer of a city is a huge attractor and is itself, a new industry. Activities that were once found only inside museums and theatres are now in the streets and spaces of cities, where one can enjoy street performance, concerts, art installation, and dance. The public realm landscape is the new stage for cultural events. This openness and generosity reflects a lively and open city where people from all parts of the globe can participate and integrate.

设计和可持续的城市
Design and Sustainable Cities

最后，让我们谈谈设计及它在可持续发展中的作用。物理形式或设计经常决定城市景观的寿命。设计所采用的形式和材料可以在很大程度上决定城市环境的成败。成功的设计将建立人和场所间的联系，赋予场所特色、记忆、身份、方向和个性。

Lastly, let's talk about design and its role in sustainability. The physical form, or design, will often determine the longevity of an urban landscape. The actual form and physical content of the design can largely determine the success of urban environments. If successful, the design will enable people to make an emotional connection to a place by imbuing it with character, memory, identity, orientation and individuality.

仅凭 “智能的”科技和可操作的生态系统便可以获得可持续性是一种虚幻的观点。人也是自然界的一部分（也是自然等式中的一个变量）。如果人不在智力和情感上投入，任何东西都不能长久存在。如果不是按照人的精神、心理和情感需要来设计的话，建造一个项目或社区所需的所有“智能的”科技、适宜的材料和能量都会是一种浪费。它不会有吸引力，寿命也很有限，但好的设计可以弥补这一不足。

It is a false belief that one can achieve sustainability based only on "smart" technologies and functioning ecosystems. People are part of the environmental equation. Nothing can sustain itself over time if people are not invested in it either intellectually or emotionally. All the smart technologies, appropriate materials and energy used to build a project or community will be wasted simply because it was not designed to the spiritual, psychological and emotional needs of people. It simply will not have appeal and therefore has a limited shelf-life. Good and great design can help to achieve this.

设计本身不能让城市成功，因为城市是复杂的运动体的组合，但城市公共空间元素的设计质量却非常重要，它可以使城市功能最大化。设计质量是一种使城市获得最大潜力的重要因素。

Design in itself cannot make cities successful, as cities are a very complex layering of moving parts. However, for a city to function maximally, the design quality of a city's public realm components becomes extremely important. Design quality is crucial factor in whether a city can reach its fullest potential.

结论
Conclusion

很明显，全球范围的城市化趋势和资源限制的现实使得我们行业最重要的作用是通过规划和设计提高城市密度，创造宜居而健康的城市环境。城市中的景观空间可以通过吸引聚集人口应对全球变暖和减少自然资源消耗。然而这只有在城市景观设计过程中超越功能需要，才能使激发灵感的、具有吸引力的场所得以实现。

It is clear with the global trend towards urbanization and the realization that we have limited global resources, the most important role we have professionally is to encourage densification through planning and designing liveable and healthy cities. A city's public realm landscape can play a crucial role in helping to combat global warming and shrinking natural resources by attracting people to them. However, this can only be achieved if our urban landscapes are designed to be more than merely functional: but as wonderful, inspired, attractive places to live and work for all socio-economic levels.

案例赏析

20 世纪 60 年代的美国大地艺术盛行，其中有很多杰出艺术家的作品都对玛莎·施瓦茨后来的创作产生了重要的影响。

图 1 是罗伯特·史密斯所创作的名为“沥青流下来了”的艺术作品。他关注了艺术以及创作的本质。他的作品反映了沥青流下来的过程。通过这个作品，他阐释了自己对地景的认识。

克里斯托弗是另一位十分著名的地景艺术家，图 2 是他的代表作：篱笆墙。在这个作品中他表达出了景观的政治属性，因为在创作中他和当地的农民进行了长时间的交涉。其实地景是一个充满政治性的整体，每个个体都有权利表达自己的意见。

通过图 3 我们可以发现，从人的视角看去，其实映入我们眼帘的大部分还是大地和景观。现在我们都认为建筑具有标示性，但是景观其实一样也可以具有标示性。

1 | 3
2 |

面包圈公园（Bagel Garden）
该公园已经存在30年了，但是玛莎·施瓦茨做的艺术展示却依然很新。这是她刚毕业时的作品，作品本身是一种政治宣言，是对以前地景的一种挑战。在这个创作中她运用了面包圈——一种犹太族食物作为景观元素。因为面包圈便宜易得，并且对环境很友好。这个创作体现出一种民主精神，她希望通过这个创作阐释景观是表达个体艺术想法的场所。

临时性景观：法式与日式两种风格的景观对比
这个作品向大家展示了一种"基因移植"的理念，她将法式园林和日式枯山水并置、叠合。这也是她第一次在城市环境中做设计。这个设计的预算紧张，并且也没有可以利用的水资源。

这是一个农场艺术作品。它的目的是与农场主进行互动，可以给农场主一些指示，帮助农场主种植、收割庄稼。当地主要种植玉米与干草。

这是一个没有什么树木的石铺地广场，在广场上用轻质的填充物做了很多地景。它象征着冰川时代冰川融化流过的痕迹。

这是在停车场上部做的一个设计，它的预算很低，并且因为上面无法种树，所以玛莎·施瓦茨做了很多具有树木象征意象的景观元素，人们可以在下面活动、乘凉。

这是玛莎·施瓦茨在设计中做的一个代表四季的盒子，这为人们提供了一个私密空间，人们在其中可以休憩、谈恋爱。

这是一些粉色的混凝土喷泉装饰柱。这其实是一个错误，本来它们并不是粉色的。而这也告诉我们其实错误本身也有可能带来更多的趣味，所以作为一个优秀的设计师，我们要学会区分好的错误和真正的错误。

这是美国亚利桑那的梅萨艺术中心设计。

它是将街区简称为艺术活动中心，而街道则是为了表演而设计的。

这里十分缺水，所以对水资源的运用十分注重效率，在设计中对水的运用进行了科学的调节，力图用最少的水造就最大化的效果。

不同尺度的空间设置让不同的活动同时进行。现在这里成为亚利桑那州最吸引人的拍摄婚纱照的地方。

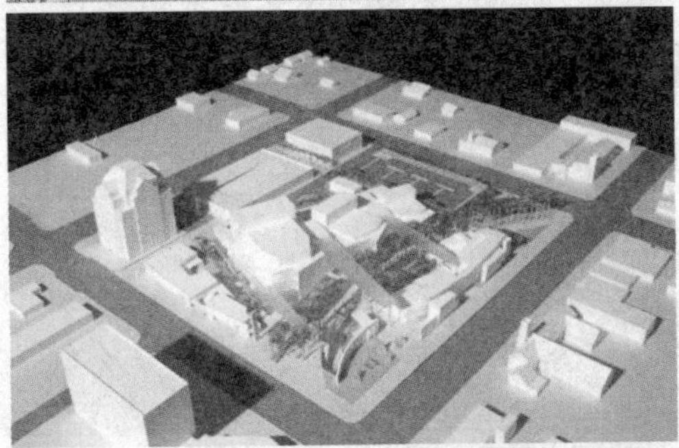

希望这里成为一个城市中的马戏场。

其实无论孩子还是大人都对水充满了兴趣。

这是玛莎·施瓦茨在芝加哥做的一个竞赛项目，是一个长3000米的老工业码头。在其中有很多的元素，如主题公园、湿地公园、沙滩、码头等。

在海水中玛莎·施瓦茨做了很多木质的伸出的小码头。

这是一个新设计的游乐场，和港口相连，人们可以在这里做游戏、游泳、玩滑板、垂钓。

多共性的场所带来了无限的可能。

图 1

图 2

图 3

图 1 是位于伦敦大象城堡的一个社区级景观设计，这里是一个地铁站。

图 2 是一个公交站的设计。

图 3 的设计为儿童带来了不同的感受，它改变了孩子们生活的场所。特别是给小男孩一个表现自我的空间。

这是位于曼彻斯特的交换广场。

作为工业革命的发源地，曼彻斯特拥有发达的铁路系统，所以设计的时候运用了很多铁路的元素。

其中布置了很多的长椅，当天气晴好的时候，这些长椅为人们提供了停留之处。

这里曾经有一条河流，在设计中也加入了河流的元素，这吸引了很多的孩子来这里玩耍。

都柏林的运河广场项目

项目通过广场的景观改造达到吸引人们来投资的目的。演艺建筑由里伯斯金设计。尽管这个场地是在建筑之前设计的，但是玛莎·施瓦茨还是运用了里伯斯金建筑中的元素，因为她认为景观和建筑是一体的。

因为这是一个演艺建筑，所以玛莎·施瓦茨将红地毯的元素也引入设计中，并利用当地植物形成一条植物带作为绿带。

现在这个设计已经成为了伦敦道克兰地区的象征，当地人都知道“红柱子”在哪里。

现在在这个项目周围的产业都发展得很好，尽管在道克兰产业的发展前景并不好。这也告诉我们景观对地区活力的提升起到了至关重要的作用。

这是为迪拜设计的一个广场，周围都是密集的街区。

玛莎·施瓦茨的设计就像是在机器上覆盖一整块布一样。其中线性的连接是为了使各个街区产生联系。这个项目中的水源都来自于建筑，在设计中也非常注重水资源的利用。中间开敞的空间将是男人们喜欢使用的空间，它比较公众化。周边这些小的具有私密性的空间则是女性的最爱。

这个项目是玛莎·施瓦茨在中国的第一个项目——中国西安世界园艺博览会迷宫园。

这个项目是位于西安的一个具有展览性质的公园。每个设计师都分到了一个 30 米 ×30 米的地块。

玛莎·施瓦茨认为西安本地寺庙等古建中拱的比例十分美，所以在这个公园的设计中也运用了许多石拱门。

在设计中还大量地运用了镜子，由于墙之间不是平行的，影像被反射后就营造出一种迷宫的氛围。

在这个“迷宫”的最后是一个玻璃房子，透过玻璃可以看见一片“树林”。

在砖墙之间种植了很多当地的柳树，当柳树长高后，柳条就自然地从墙上伸出垂下，营造自然之美。

通过砖墙、石拱、镜子和柳树，玛莎·施瓦茨在城市中营造出一个奇幻的谜院，人们喜怒哀乐的真实情感毫无保留地在游园过程中展现出来。单向透视镜一面反光成像，而另一面可以透视对方。这样出园的人们可以悄无声息地在暗处观察游人的行为和举动。人们体验到了从城市空间到自然空间的过渡，并且在镜子的反射中体味着“看与被看”的乐趣。这也很好地契合了“城市与自然和谐共存”这一设计主题。

所有图片均来自玛莎·施瓦茨 2013 年 4 月天津大学演讲

图书在版编目（CIP）数据

玛莎·施瓦茨 & 天津大学建筑学院风景园林系联合设计教学实验 / 曹磊主编. -- 南京 : 江苏凤凰科学技术出版社, 2015.3
ISBN 978-7-5537-2624-3

Ⅰ. ①玛… Ⅱ. ①曹… Ⅲ. ①园林设计－教学实验－高等学校 Ⅳ. ① TU986.2-45

中国版本图书馆 CIP 数据核字 (2015) 第 035864 号

玛莎·施瓦茨 & 天津大学建筑学院风景园林系联合设计教学实验

主　　编　曹　磊
项目策划　凤凰空间/高雅婷
责任编辑　刘屹立
特约编辑　陈丽新

出版发行　凤凰出版传媒股份有限公司
　　　　　江苏凤凰科学技术出版社
出版社地址　南京市湖南路1号A楼，邮编：210009
出版社网址　http://www.pspress.cn
总　经　销　天津凤凰空间文化传媒有限公司
总经销网址　http://www.ifengspace.cn
经　　销　全国新华书店
印　　刷　北京博海升彩色印刷有限公司

开　　本　787 mm×1 092 mm　1/12
印　　张　8
字　　数　71 000
版　　次　2015年3月第1版
印　　次　2024年4月第2次印刷

标准书号　ISBN 978-7-5537-2624-3
定　　价　35.00元